SOIL CONSERVATION, MANAGEMENT AND ANALYSIS

About the Author

Dr. Madireddi V. Subba Rao got his M.Sc. Zoology with I Class and I Rank in 1963 from Agra University, Agra; awarded Ph.D. for his studies on 'Eco-Physiology of two selected lizards' in 1968 from Sri Venkateshwara University, Tirupati. He was Post-Doctoral Research Fellow at Creedmoor State Hospital, Queens, New York, USA in 1969 and D.Sc. (IOU, Sri Lanka) in 1997.

He was Visiting Professor in Biology, Hofstra University, Hempstead, L.I., New York, U.S.A. (1969) . He worked as a Senior Professor, Chairman and Coordinator, Department of Environmental Sciences, Andhra University, Visakhapatnam, India and retired in 2001. Now he is a President in Environmental Research Academy, International (EnRA).

In recognition of his achievements in contemporary research, National Environmental Science Academy, New Delhi awarded him a Best Scientist for the year 1986 and decorated him with a Gold Medal and in 1991, the Andhra University recognised the research talents of Prof. Rao and awarded him the Best Research Scientist and ISEAS, Bangalore with Gold Medal in 1997, American Biographical Research Inc., USA has nominated him as International Distinguished Leader in 1996, 1997. His name was placed in Asia's Who's Who, Man of Achievement, 1996, 1997, 2000. He was awarded a gold medal for best Zoologist by ZSI in 2004 AZRA and AZra in 2010. NTPC, Vamsadhara, YRP, Visakhapatnam Port Trust, NFCL, GFCL, Spectrum, Kakinada Sea Ports Ltd., etc.

SOIL CONSERVATION, MANAGEMENT AND ANALYSIS

Prof. Madireddi V. Subba Rao R, *Ph.D.*

2014

Daya Publishing House®

A Division of

Astral International Pvt. Ltd.

New Delhi – 110 002

ISBN 9789351301233

Published by : **Daya Publishing House®**
A Division of
Astral International Pvt. Ltd.
– ISO 9001:2008 Certified Company –
4760-61/23, Ansari Road, Darya Ganj
New Delhi-110 002
Ph. 011-43549197, 23278134
E-mail: info@astralint.com
Website: www.astralint.com

Laser Typesetting : **Classic Computer Services**
Delhi - 110 035

Printed at : **Thomson Press India Limited**

PRINTED IN INDIA

Dedicated to
my Beloved Betterhalf

Mrs. Madireddi Ratna Kumari, M.A. (AU)

for her excellent encouragement and Interest

FOREWORD

Today is the 39th World Environment Day. The main objective is to "Conserve Our Natural Resources and to protect our Enviroment". The soil is one of the natural resources and a part of the environment.

The United States Departments of Agriculture (1951) defined soil as the "Collection of Natural bodies occupying portions of the earth's surface and that support the living organisms. The earths crust is the origin of soil. On the process of weathering, primary rocks produces which is known as "Soil Parent Material" and the rocks from which this material is produced called "Parent Rocks".

Soil contains nutrients and minerals. Plants need only nutrients. The mineral particles in soil are distinguished by its size. Sand is the largest than silt and the smallest are clay. The proportion of each of these components give a soil and its particular characteristics. A soil with a lot of sand and little dry with many air spaces and drains early. But it have less nutrients. A soil with mostly clay particles holds more water and is slower to drain. A loamy soil with balanced amounts of sand, silt and clay is best suited for agricultural use.

Life on the earth depends upon the living soil. Most of the food of the world is grown in soil. Soil supports the growing plants and serves as a reservoir of water, air and nutrients which are essential for living plants. It is significant that we take care of soil so that the soil can take care of us and other life on earth.

In view of the above emerging situations, the concern over the quality of soil has gained paramount importance for susceptibility of mankind. Soil is a combination of various parameters like Physical, Chemical, Biological and Bacteriological characteristics.

Ministry of Environment and Forests has critically studied for ascertain the allowable limits of all the above parameters and fixed certain norms for implementation of land under Land Act (1994) and Environmental Protection Act (1986). In order to quantity of the parameters, the standard methods have been evaluated and supported by recent development in instrumentation.

The author of this book on 'Soil Conservation, Management Analysis' made an effort with his vast experience in this line to bring various analytical techniques/methods together for evaluation of soil quality parameters. This book contains very relevant information on this valuable subject and can be useful for the Post & Under Graduates, Researchers, Scientists, Municipalities, Local bodies, Hospitals, Agriculturalists, Pollution Control Boards and Industries for information searching processes.

I hearty congratulate to Prof. Madireddi V. Subba Rao, Retd., Sr. Professor, Head and Co-ordinator, Department of Environmental Sciences, Andhra University, the author of this monograph for his sincere efforts in building an house of information in respect of Soil Conservation, Management and Soil Analysis. I am very sure that this type of publication will be a source of inspiration and awareness among Academicians, Agriculturists, Scientists, Engineers and Common public.

New Delhi

5th June 2012 **Prof. S.M. Mukherjee**

PREFACE

The US Department of Agriculture (1951) defined soil as the "Collection of National Bodies Occupying Portions of Earths Crust and that support the living organisms. The Earth's crust is the origin of the soil on the process of weathering, primary rocks produces which is known as "Soil Parent Material" and the rocks from which this material is produced known as "Parent Rocks" (Chapter 1).

Life on the Earth depending upon the living soil. Most of the food is grown in Soil. Soil supports the growing plants and serves as a reservoir for water, air, nutrients and minerals, which are essential for living plants. It is significant that we take care of soil, so that the soil can take care of us and other life on earth.

In view of the above emerging situations, the soil is a combination of various parameters like Biological, Physical, Chemical and Bacteriological Characteristics.

Recent researches on soil reveals that 75 billion tonnes of top soil (surface soil) are eroded (soil erosion) (Chapter 4) away every year by different geological agents and converted into waste land (Chapter 2). New soil is formed (soil formation) (Chapter 1) from parent rocks, one cubic meter of soil may be formed from 100 to 1000 years to build up. Soil is no longer a renewable resources (Chapter 3). Only top layer of this soil with humus (wet) (Chapter 2) is vital to grow food producing fertitle land is lost by desertification (Chapter 5). The rate of desertification is exponential. It was covered by forests (Chapter 9). Every year, 60,000 sq. km. of productive land are lost which affects 1000,000 million people world wide.

Having realized the threat of the soil resources, the entire world community is uniting today for preparation of drafts for conservation measures (Rio-Earth Summit, 1992; Stockholm Conference, 1972; G-7 Summit, 1989 etc.) (Chapter 8). Every minute, chunk of fertile agriculture land of the size of "Tennis court" is being converted into waste land (Chapter 2). Thus loss of soil by soil erosion (Chapter 4) is an age old problem. To challenge this problem, soil conservation techniques are being applied (Chapters 6 & 9). Soil conservation is the protection of soil against excessive loss of fertility (Chapter 7) by natural, artificial, biological and chemical means. All of us can take part in soil conservation process. These are many ways that each of us can help in solving this environmental education monitoring and soil management (Chapter 10).

My sincere thanks are due to my wife, Mrs. Madireddi Ratnakumari, my children my children, Sunil, Dr. Madhavi Shiva and my grand children, shriya, Rhea, Sai Suhaas, Sai Sanjit, Sai Darhaas and Sai Sanvita for their interest and encoouragement.

Finally, I thank Mr. Anil Mittal of Daya Publishing House, Ansari Road, Daryaganj, New Delhi for their neat execution and printing of the second edition of this book in time.

The Author welcomes any suggestions, criticisms regarding the contents of this book that would help to revise this Monograph and make this more applicable.

Prof. Madireddi V. Subba Rao

Visakhapatnam, A.P.

E-mail: musubbarao_enra@yahoo.co.in

Contents

CHAPTER 1

Formation of Land

INTRODUCTION

India extending over a 329 million hectares has one of the richest Biological diversity due to its diverse physical and climatic factors. India is one of the 12th identical mega-diversity centres. Out of 329 mha of land is urban and other non agricultural use; 21 mha is barren land. Only 266 mha of land is available for potential use. This potential land in India is divisible into 143 mha and agricultural land and this agricultural land have suffers from 83 mha for soil degradation. Out of 40 mha of land is completely unproductive and 83 mha is partly productive land.

India's protected area includes 75 National Parks and 421 sancturies covering an area of 1,40,675.46 sq.km.

According to India's National Forest Policy (1952), 33% land should be under forest cover but, at present, the area under forest cover in India is only 14%. India is losing 1.5 million hectares (mha), 74 mha to 40 mha. In India, tropical forests cover about 58.8 mha and constitute about 9% of the country's forest cover. Majority of these are endemic, rare and threatened plant and animal species. They are also home for several rare tribes and forest dwellers. Now forests are seriously threatened today. Most area have been cleared. Ecological management of soils and the forests are very important for conservation purposes and sustainable development of land use in India.

LAND FORMS

Land forms, the distinctive geometric configuration of the earth's land surface, are of primary concern to is because they exert far-reaching and fundamental influence on the patterns of human activity. The systematic study of the earth's relief features is known as geomorphology (geo-earth morph, formology, science), but for simplicity may well be called ("the study of land forms"), landscape features must be sorted out into classes or groups of essentially similar both in outward form and in origin. The Geomorphologist is equally interested in forms and in the processes and stages of development at these forms. Landscape features pass through an orderly series of changes, just as do human beings in their life span. Once these life stages are known any landscape feature can be related to a particular event in the life cycle of land forms, thereby lending order and natural low to out concepts of land form development. It is possible, of course, to describe all landscape features by tabulating their size, shape, angles of slope, and orientation without thought to their origin or development.

In terms of the grandscale of geological processes, there are two .fundamental classes of land forms. First, there are the original crusta! masses raised by internal earth forces and by volcanic eruptions. These comprise the initial land forms; second there are the land forms made by agents of denudation. Because these follow the initial forms and occur in orderly sequences they represents nothing less than the existing stage in a great struggle or contest. The internal earth forces spasmodically elevate parts of the .crust to create initial land forms. The external agents patiently keep wearing these masses down and carving them into vast numbers of smaller sequential land forms. Sequential land forms are products of one or more of the land scralpturing agents running water, waves, ice and wind. These erosional agents, aided by processes of rock decay and downside movements of sol! and rock under gravitational force, attack from the outside all continental rock masses that become elevated by mountain-making unheavafs or wrapping of the earth's solid crust. No part of the land surface is ammine from attack. Examination of a freshly out cliff, such as that in a new highway cut or quarry wall, will reveal several kinds of earth materials. Solid, hard rock which is still in place and relatively unchanged is called bedrock. It grades upward into a zone where the rock has become decayed and has disintegrated into clay, silt and sand particles. This may be called the weathered over burden, or residual over burden. At the top is a layer of free soil, often called top soil by farmers and gardeners. Over the soil may be a protective layer of grass, shrub or trees. The first step in approaching geomorphology therefore, is to learn the fundamental facts about rocks of the earth's crust, their composition, physical and chemical properties, characteristic forms of occurrence, processes or origin, and geologic age. A knowledge of common rocks and the predominent minerals that comprise them is of further value to us because rocks and minerals have intrinsic value to man as ores, fuels, and construction material's. Rocks may divided into three major classes 1) ingeous 2) sedimentary and 3) metomorphic according to their origin. Igneous rocks solidify from a molten state. The highly heated fluid rock, termed magma, originates at considerable depths beneath the earth's surface and is forced by great internal pressures to prenetrate the solid, brittle outer crust of the earth. Those rocks that are composed of particles derived from previously existing rock, then laid down after transportation by streams waves or tidal currents, wind or ice are sedimentary rocks. Rocks of any origin may be the parent material for sediment in addition, sediment may result from chemical reaction and precipitation. As the solid particles settle out, a layer is formed. Sedimentary rocks are distinguished from igneous rocks by the usual presence of layers resulting from changing size and composition of sediment, the layers are called strata or simply beds. The planes of separation between successive layers are planes of stratification. The rock is said to be stratified, any of the types of igneous or sedimentary rocks may be altered by the tremendous pressures and high temperatures that .accompany mountain building movements of the earth's crust. The result is a rock so greatly changed in appearance and composition as to be classified as a metamorphic rock. Generally speaking, metamorphic rockers are harder and more compact than their original types, except when the latter are igneous rocks. Repeatedly writings on geology, reference is made to a fault, or surface of breakage in the earth's crust along which a slipping movement has taken place one mass sliding past the other. Faults are of great importance geologically because they result in unlike rock types being brought into contact at the same level. Many layered rocks, such as sedimentary strata or leaves have experienced compressional forces in the earth's crust which bend the layers into wave like corrugations termed folds. This happens not only on a vast scale of mountain dimensions, but also on smaller scales down to tiny wrinkles a fraction of an inch from crust to crust. Often, as in maryle scale and slate, many scales of folds are super imposed in the same rock mass. Large folds have a strong influence in shapping land forms.

LANDFORMS MADE BY STREAMS

Streams erode in various waves, depending on the nature of the channel materials and the tools with which the current is armed. The force of the flowing water above, exerting impact and a dragging action upon the bed can erode poorly consolidated alluvial materials such as hard, gravel, silt, and clay, a process termed hydraulic action wherein rock particles carried by the swift current strike against bed rock channel walls, chips of rocks are detached. The rolling of cobbles and boulders over the stream bed will further crush and find smaller grains to produce an assortment of grain sizes. The load of a stream is carried in three forms, dissolved matter is transported invisibly in the form of chemicals sons. All streams carry some dissolved salts resulting from rock decomposition. Clay and silt are carried in suspension, that is, held up in the water by the upward elements of flow in turbulent eddies in the stream. The load carried by a large river is of considerable importance in planning for construction of canal system for irrigation. Sediment will be trapped in the reservoir behind a dam, eventually filling the entire basin and ending the useful life of the reservoir as a storage body. We tend to think of a river in flood as changing largely through increase in height of water surface, which causes channel overflow and laundation of the adjoining flood plain because of the turbidity of the water. We cannot see the changes taking place on the stream bed, but these can be determined by sounding the river depth during stream gauging measurements.

LIFE HISTORY OF A STREAM

Throughout its life history, a stream passes through a series of stages, each with certain definite characteristics. The initial stage occurs as soon as a new land surface is created by uplift and dislocation of a portion of the earth's crust. It is assumed here, for simplicity of discussion, that the surface was formerly under ocean level and has now become exposed for the first time. The land scape is thus composed entirely of initial sand forms. Rain falling upon the land will produce over land flow. This must flow down the initial slopes, whatever their form. Water flow will be concentrated where slight depression exist in the slopes, thus causing the development of stream channels, which are quickly depended by erosive action of the water and lose rock particles as it carries. Once formed, the stream enters upon the stage of youth. Depending of the channel is the principal activity of a young stream, whose capacity for load exceeds the load available to it. Lake outlets are cut through, draining the lakes and extending the stream across the lake floors. Waterfalls are cut down at the lip until they are nothing more than rapids. A depending gorge or canyon is perhaps the most striking land form associated with a young stream. The stage of maturity is reached when the stream has completed its phase of rapid down cutting and has prepared itself a smoothly graded course. Immediate evidence of the earliest stage of maturity is the beginning of development of a float valley floor. During enlargement of a bend the river channel shifts towards the outer part of the bend leaving a strip of the relatively flat land or floor plain, on the inner side of the bend. The deposit of clay, sand or gravel made by a stream where it flows into a body of standing water is known as delta. Deposition is caused by rapid reduction in velocity of the stream current as it pushes out into the standing water. The course particles settle out first, the fine clays continue out farthest and eventually come to rest in fairly deep water. Deltas show a variety of shapes. The Nile delta whose resemblance to the Greek letter "delta" suggested the name for this type of land forms, has many distributions which branch out is a radial arrangement. Deltas of large rivers have been of great geographic importance from earliest histroical times because their extensive flat areas support dense agricultural population.

Consider now the entire aspect of denudation of a large region under the combined attack of weathering, mass, wasting, overland flow and streams. Imagine an area, such as a continent or a large portion of a continent, which is elevated by internal earth movements to provide a new land mass. This elevation will constitute the initial stage of a grand cycle of land mass denudation, in which the region passes through youth, maturity and old age. For the sake of simplicity, it must be assumed that the elevation of the land mass occurs rapidly and that further crustal deformation then causes, leaving the denudation agents a long period of uninterrupted activity.

This for this type of land forms, has many distributions which branch out is a radial arrangement. Deltas of large rivers have been of great geographic importance from ealiest histroical times because their extensive flat areas support dense agricultural population.

Consider now the entire aspect of denudation of a large region under the combined attack of weathering, mass, wasting, overland flow and streams. Imagine an area, such as a continent or a large portion of a continent, which is elevated by internal earth movements to provide a new land mass. This elevation will constitute the initial stage of a grand cycle of land mass denudation, in which the region passes through youth, maturity and old age. For the sake of simplicity, it must be assumed that the elevation of the land mass occurs rapidly and that further crustal deformation then causes, leaving the denudation agents a long period of uninterrupted activity.

LAND FORMS MADE BY WAVES AND CURRENTS

Ocean waves work unceasingly to transform the shores of the continents and lands. Ocean waves and ocean currents generated by the wind. Waves travel across the deep ocean with only gradual loss of energy, but when shallow water is reached, waves form changes radically and new water motions are developed. These take the form of powerful surge's and currents capable .of performing great erosive and transportational work'. Most shortlines have a rather smooth, sloping bottom extending out beneath the water level. As waves approach this shallow zone, wave velocity becomes less and wave crusts become more closely spaced. Wave height and steepness increase rapidly until the, crest roils forward to make a breaker. After the breaker has collasped, a foamy turbulent sheet of water rises up the beach slope. This swash, or uprush is a powerful surge, causing a landward movement of sand and gravel on the beach.

In times of relative calm or moderate winds, waves do little erosional work but tend in tend build beaches and other deposits out of sand and gravel. In times of storm, when enormous waves break and throw tons of water against the shoreline, erosion is rapid. The violent uprush hurls cobbles and boulders against exposed bedrock along the shore, causing fragments to be broken free. A continual crushing and grinding action goes on as the stones are jostled together. The products of this breaking up are sorted out according to the to size. Fine particles are carried out to sea. Where they eventually come to rest in deep, quiet water to form clay and silt layers. Sands and gravels remain close to shore, forming beaches and bars, the abound flood currents generated by tides perform several important functions along a shoreline. First, the currents that flow into and out of bays through narrow inlets are very swift and can scour the inlet strongly to keep it open despite the tendency of shore drifting processes to close the inlet with sand. Second, the tidal currents carry much line silt and clay in suspension, derived from erosion of marine clif's from streams that enter the bays or from bottom muds agitated by storm wave action. This fine sediment tends to become clotted into small aggregates where fresh water mixes with salt water. The sediment that settles to the floors of the bays and estuaries where it accumulates in layers and gradually fills the bays and much organic matter is normally present in such sediment.

Saltrnarsh is of interest to geographers because it is land that can be drained and made agriculturally productive. The salt marsh is first cut off from the sea by construction of an embankment of earth, a dike, in which gates are installed to allow the fresh water drainage of the land to exist in ebb flow but to prevent the increase of salt water during flood flow, gradually the salt water is excluded and soil water of the disliked land becomes fresh. Such diked lands are intensively developed in Holland and southeast England.

SHORELINE CLASSIFICATION

Because of shore line display a wide variety of forms, it is desirable to have a classification to group them according to their origin and development. Five major classes of shorelines have been found to include five known types:

1. **Shoreline of Submergence:** Wherever a sinking down of the earth's crust occurs near the border of a land area, or there is a worldwide rise to sea level, the new shoreline

takes a position along what was approximated a former contour of the land surface. Below this level, all the surface formly exposed to the air is now submerged beneath the sea. In this way, the term shoreline of submergence is applicable.

2. **Shoreline of Emergence :** Wherever a rising of the earth's crust has occurred near the boarder of a continent, or th sea level has fallen, a shoreline of emergence is created. The water line takes a position against what was formely a slope of the sea floor. Above the new shoreline lies a new coastal land belt which has emerged from the sea. Withdrawal of water to form extensive continental ice sheets is an effective cause of the sealevel lowering. It is likely that shorelines of emergence were widely distributed during the time of maximum glacier advance in the Pleistocene epoch.

3. **Neutral Shorelines :** Wherever a shoreline has been : formed as a result of new materials being built out into the water, it is classified as a neutral shoreline. The word "Neutral" implies that there is no requirement of a relative change between the level of the sea and the coastal region of the continent.

   Several types of Neutral shorelines are known today. Each type is easily understood when referred to the agent responsible for building out the material into the water. An alluvial fan shoreline curved in outline, is built in braided streams building a fan. Very similar shorelines are formed by building of a glacial outwash plain into the sea where the ice front stood near the shore. A delta shoreline is built of material brought out by a stream system. Where a volcano eruption has occurred, the slope of a volcano or the edge of a lake flow may compose the shore, which is then classified as a volcano shoreline. A coral reef shoreline is built by marine organisms in the shallow water zones of tropical seas.

4. **Fault Shorelines :** An unusual type of shoreline is produced by faulting or the earth's crust in such a way that there is

dropping down of a segment of the crust on the seaward side of a break and corresponding raising up of the landword side, should the down dropped block subside below sea level, the shoreline will come to rest against the steeply sloping land surface which is the surface off the plane of faulting and can, be termed a fault shoreline. A similar result would be obtained where a crustal mass bounded by fault planes arises from beneath the sea.

5. **Compound shorelines:** These are those that show the forms of two of the previous classes combined, for example, submergence followed by emergence, or vice versa.

LAND FORMS MADE BY WIND

Wind one of the agents of erosion, produces a variety of interesting sequential land forms, both erosional and depositional. Wind performs two kinds of erosional work. Loose particles lying upon the ground surface may be lifted into the air or rolled along the ground. This process is deflation. Where the wind drives sand and dust particles against an exposed rock or soil surface, causing it to be worn away by the impact of the particles, the process is abrasion. Abrasion requires cutting tools carried by the wind, deflation is accomplished by air currents alone.

Deflation acts wherever the ground surface is thoroughly dried out and is littered with small, loose particles derived by rock weathering or previously deposited by running water, ice, or waves. Thus, dry river causes, beaches, and area of recently formed glacier deposits are highly susceptible to deflation. In dry climates, virtually the entire ground surface is subject to deflation because the soil or rock is everywhere bare. Wind is selective in its deflational action. The finest particles, those which constitute clay, or silt, are lifted most easily and raised high into the air. Sand grains are moved only by moderately strong winds and tend to travel close to the ground. Gravel fragments and rounded pebbles upto 2 to 3 inches (5 to 8 cm) in diameter may be rolled over flat ground by strong wind but do not travel far. They become easily lodged in hallows or between their follows. The principal land forms produced by deflation is a shadow depression termed blow out, or deflation hollow. The depression may be from a few yards to a mile or more in diameter, but is usually only a few feet deep, blowouts form in plains regions in dry climates. The sand blast action of wind against exposed rock surfaces is limited to the basal few feet of a clif, hill, or other rock mass rising above a relatively flat plain, because the grains do not rise high into the air. Wind abrasion produces pits, grooves, and hallows in the rock. Where a small rock mass projects above the plain, it may be cut away at the base to make a pedestal rock, delicately balanced upon a thin stem. Most pedestal rocks, or mush room rocks, however, are produced by weathering processes."

DUST STORMS AND SAND STORMS

In dry seasons and over plains regions, strong, turbulent winds like great quantities of fine dust into the air, storming a dense, high cloud called a dust storm. The dust storm is generated where ground surfaces have been stripped of protective plant cover by cultivation or grazing or where they naturally carry no vegetation cover because of extreme aridity of the climate. It has been estimated that as much as 4000 tons of dust may be suspended in a cubic mile of air (875 metric tons per cubic kilometer). On this basis, a dust storm 300 miles (500 km) in diameter might be carrying more than 100 million tons (90 million metric tons) of dust enough to make a hill 100 feet high and 2 miles (3 km) across the base. A region that supplied the dust for thousands of such storms would thus lose a considerable mass over a span of thousands of years. Whether the same period streams would remove more material from the same area is difficult to say, but in all probability they would remove much more. The erosional effect of blown sand is concentrated on surface exposed less than a foot or two (0.3 to 0.6 m) above the flat ground surface.

SAND DUNES

A dune is any hill or accumulation of sand shaped by the wind. Dunes may be active or live where bare of vegetation and constantly changing from under wind area. They may be inactive, or fixed down covered by vegetation that has taken root and serves to prevent further shifting of the sand. Several common varieties of dunes are treated here. The crescent dune or barchan, is an isolated dune, which in plain view resembles a blunted erescent. The broadly rounded ends of the erescent point down wind and indicate the direction of dunemotion and prevailing winds. Where sand is so abudant that it completely covers the ground, dunes take the form of wave like ridges separated by trough like furrow.. The dunes are called transverse dunes because their crests trend at right angles of direction of wind.' The entire area may be called a sand sea, for it resembles a strom rosed suddently frozen to immobility. The term erg, referring to any large expense of dunes in the Sahara Desert, has been adopted by geographers for this type of land scape. Another group of dunes belong to a family in which the dune crest is bowed conversly downwind, the curve of the curvature of crests in the barchan and transverse dunes. These may be described a parabolic in form. In arid plains and plateau, where vegetation is sparse and winds strong, groups of parabolic blow out dunes develop to be of shallow deflational hollows. Sand is caught by low bushes and accumulates in a broad, low ridge. These dunes have no steep slip faces and may remain relatively immobile. In some cases, however, the dune ridges migrates down wind, drawing the parabola into a long narrow form with parallel sides. This form resembles a hair pin in plain, hence has been called a Hair pin dunes although it is a member of parabolic family.

Still another class of dunes is described as longitudinal because the dune ridges run parallel with the wind direction. On desert plains and plateaus, where sand supply is meager but winds are strong from one direction, longitudinal dune ridges are formed. These are only a few feet high, but may be several miles long. Longitudinal dune ridges oriented parallel with dominant winds, occupy vast areas of central Australia referred to as Sand ridge deserts.

In several parts of the world, the ground is underlain by deposits of wind transport, silt which has settled out from dust storms overs many transcends of years. The material thus formed is known as loess. It generally has a uniform buff colour and commonly lacks any visible layering or other landing. Loess has a tendency of break or cleave, along vertical cliffs, wherever it is exposed by the cutting of a stream or by man. The chavage is possibly produced by a slight shrinkage of the entire mass as it has compacted after being laid down. The importance of loess in world agricultural resources cannot be easily overestimated. Loess plains and plateaus have developed rich, black soils especially suited to cultivation of grains. The highly productive plain of southern Russia, Argentina, Pampa and the rich grain region of north China are underlain by loess. Because loess forms are vertical walls along valley sides and is able to resist sliding or flowage, but at the same time is easily dug into, it has been widely used for cave dwellings both in China and in Central Europe.

CHAPTER 2

Waste and Wet Land Management

A WASTE LAND MANAGEMENT

Waste, year after year, decade after decade, more and more is generated, and as the precious time progresses, precious space for it decreases. But some communities have clearly identified this problem and they are saving space and money and therefore, saving the environment as well. Waste management is a vital component of the environmental movement. Every one contributes to the problem therefore; everyone can help contribute to the solution. That solution is simply reducing, reusing and recycling.

Waste is unavoidable by product of human activities. Since the beginning humankind has been generating waste, be it bones and other parts of animals they slaughtered for their food, or wood they cut to make carts. With the progress of civilization the waste generation becomes more and more complex in nature. At the end of 19th century, the industrial revolution saw the rise of world consumers. Not only did the air get more and more polluted but the earth itself became more polluted with the generation of waste. Today we live in a 'throwaway society' because we regularly purchase things which are manufactured, purchased and used only once, then disposed off with no wary or care. We had relied on the 'out of sight, out of mind' tactics for many years, but the time has arrived to treat our waste intelligently and properly.

In times long past, when, the world population and the amount of waste produced was much smaller, the earth used to absorb the waste and we never had to give a second thought to it. That has completely changed. The population is more than 1 million population have increased from 12 in 1981 to 24 in 1991 and 35 in 2001. The smaller cities (tier I and tier II towns) have also experienced rapid population growth (Census of India, 2011).

During the seventies when India's urban population was 109 million, the per capita solid waste generation was between 150 to 350 grams per day, in nineties when the urban population was 217 million, the per capita solid waste generation was between 320 to 550 grams. Whereas, presently when the urban population is 285 million, the per capita solid waste generation is between 350 to 1,000 grams per day. It has been estimated that from 1991 to 2011 while the urban population of India may double (from 217 million to 395 millions) the quantity of waste generated may be triple (from 20.7, million tonnes to 50.3 million tonnes Bhide, A.D., 1999). In eighties, the family waste used to be 7 kg but now it is more than 50 kg. The per capita waste generation in urban areas are also varies according to size of population. In urban areas with a population of less than one million people, the per capita waste generation is 0.21 kg per day while in areas with a population of more than 5 million, the waste generated goes upto 0.50 kg per day per capita (NEERI, 2000). This clearly indicated that growth in municipal solid waste in our urban centers has outpaced the population growth in recent years. This trend can be ascribed to our changing life styles, food habits and improving living standards. In India about 22,900 million liters of waste water is generated per day. Of this only 26 per cent is subjected to any kind of treatment and almost 11,000 million liters of untreated waste water reaches the Indian rivers.

Indian cities have a striking similarity when it comes to heaps of garbage, overflowing waste bins and drains. An effective and sustainable waste management system is still rare in India. The waste in cities is managed by the respective municipalities — they collect and transport it to disposal sites. The limited revenues earmarked for the municipalities make them ill equipped to provide for high costs involved in the collection, storage and proper disposal.

As a result, a substantial part of the municipal solid waste generated remains uncollected and grows in heaps at poorly maintained collection centers. The choice of a disposal site also is more a matter of what is available than what is suitable. The average collection efficiency for municipal solid waste in Indian cities is about 72.5 per cent and around 70 per cent of the cities lack adequate waste transport capacities (TERI, 1998). The unsanitary methods adopted for disposal is therefore a serious health concern. The poorly maintained landfill sites are prone to ground water contamination. Open dumping of garbage facilitates, the breeding for disease vectors such as flies, mosquitoes, cockroaches, rats and other pests. The scenario is bad in Indian cities and nothing can be done unless the municipalities themselves wake up for it. The municipalities therefore, face the challenge of reinforcing their available infrastructure for efficient waste management and ensuring scientific disposal, by generating enough revenues either from generators or from activities that generate resource. A clean city is not an accident but is concerted effort of the citizens, the city managers and the society.

OBJECTIVES

The problem of waste, however, is not just limited to large cities. It has also seeped into small / medium cities. These cities also have in plenty of overloaded bins and dump yards and overflowing choked drains. Earlier much work has been done on waste management in larger cities (WHO 1950; Mishra R.P., 1970; Bhide *et al.*, 1973; Hussain, S.K 1974; Park, J.E. and Park, K., 1979; Jorgenson, S.E., Johson I., 1981; AM, M.I., 1993; Bhattacharya, S.C., 1997, Malik, J 1997; Rahman, A. 1998; Srivastava, N., 1999, Sanyal, 2001; Siddiqui, Z.H., 2003; Khan, N., 2005) but little or no attention was given, to smaller towns which have been described as 'temples of modern India'. Keeping the aspects in mind, Aligarh city has been selected as a study area. This study includes the evaluation of both types of waste generated from households,

(i) solid waste and

(ii) waste water

The focus of the book entitled 'Household Level Waste Management in an urban area is to conduct in depth investigations of households belonging to different income groups from the different wards of the city; for examining the nature, amount, storage and disposal of household solid waste and waste water; for examining the physical and chemical characteristics of waste; for addressing the waste associated problems like pest problem inside the house, waterlogging problem in the neighbourhoods and health problems of the residents; identifying the vulnerable households;

mapping of vulnerable areas of the city for management and planning and finally to suggest suitable measures for management of waste in the city.

WET LAND MANAGEMENT

A large part of country's land area is not yielding the desired aspects of production due to different environmental parameters and natural calamities, in addition to the area under irrigation large chunk of area is either water logged major part of time or uncultivable due to physical and geographic conditions. The area prone to the inundation during floods and due to the close proximity of lakes and reservoirs exhibit the characteristics of semi aquatic characters while the land prone for regular erosions, rocky out of crops, degraded forests and the area under snow constitutes the waste land which is partly culturable.

WET LANDS OF INDIA

Wet lands means the different lands to different people. Some think that they are water logged areas or bogs which are full of pests and weeds and should be drain out. Others think that these shallow beautiful area are natural harbours of variety of fauna and flora. Infact these areas are transitional zones between land and water showing up peculiar characteristics and inhabitants. The wet lands are natural and man made, permanent or temporary.

The international agencies like International Union for Conservation of Nature and Natural Resources (IUCNNR) and UNESCO high lighted the importance due to the significant role in biological productivity, flood control, recharge of aquifers regulation of water quality, treatment of waste water erosion control and wild life presence.

Wet lands occurs in almost all climatic regions ranging from the frozen, glaciated north to the arid tropics and differ widely in their biotic and abiotic factors. Wet lands in the modern society are drained out or filled to satisfy the ever increasing demand for the land for residential and industrial purposes, The information available on tropical wet lands is meagre, scanty and fragmentary to have a comprehensive understanding of the ecosystem.

It is estimated that out of the total geographic area of 328 million ha. around 1 50m ha are effected by erosion winder water, about 7 millions ha by water logging and salinity and another 20 million ha by floods. The area cultivate by rice is about 400 lakh ha and jute about 4.5 lakh ha which may also be considered as wet lands for all practical purposes.

As proper classification, Wet lands of India is not available. The indiscriminate deforestation resulted in sediment movement and caused incalculable damage to wet lands. Many marshes have been drained and mangroves were destroyed for wood and water bodies were filled or reclaimed for residential and Industrial purposes. Wet lands were converted as fish ponds or for cultivation of paddy or sugarcane.

WET LANDS

According to International Union of Conservation of Nature and Natural Resources (IUCNNR) defined wet lands which cover ecosystems like marshes, swamps, bogs, rivers, spring lakes, reservoirs and coastal areas as follows:

"Submerged or water saturated lands, both natural and man made, permanent or temporary with water that is static or flowing, fresh, brackish or salt is including areas of marine water. The depth of which at low tide, does not exceed 6m. The International Biological Programme (1972) gave an ecological definition is as follows:

"Part of the surrounding ecological structure and several stages in the succession from open water to dryland or vice versa, occurring at sites situated as a rule between the highest and lowest water levels as long as the flooding or water logging of the soils is of substantial ecological significance".

Cowardin *et. al.*, (1979) defined wet lands as:

"Lands transitional between terrestrial and aquatic systems when the water table is usually at or near the surface or the land is covered by shallow water".

Thus according to this definition, wet lands should show one or more characteristics of the following:

1. It should be a shallow water body that covers land with water for some time,
2. It should be and support by hydrophytes seasonally
3. The substrate is watery.

Loves and Milone (1984) defined wet lands as areas where for part of the year at least water stands naturally from 2.5 cm to 317cm.

Brij Gopal (2003) considered wet lands as the weed growing lands.

Wet Land Areas

The wet lands are of two types: 1. Natural 2. Artificial. The natural wet land bodies include fresh and salt water marshes, swamps or high rainfall regions, peats, bogs of USA and Northern Europe and the forested wet lands while the artificial wet lands which are managed by man such as wildlife sanctuaries, (*e.g.* : Ghana, Bharatpur *etc.*), rice fields and fish ponds.

Importance of Wet Lands

The developing countries like India have not yet understood their importance. This may be due to absence of information or lack of knowledge about the importance. It is now essential to understand and take up a comprehensive study of them including sociological, economic and ecological components.

Wet lands are highly productive areas acting as breeding grounds to a variety of animals and plants. They also give shelter to migratory birds specially for water fowl. They serve as excellent areas for consideration of resident and migratory species.

Wet lands in tropical areas are used to cultivate life sustain in food such as rice, cereals and economic species such as jute and sugarcane. They also support of the life of shell fisheries, prawns and fishes. Wet lands are also to serve as fish tanks. Some of them to provide grounds for recreational purposes.

Wet lands receive large quantities of storm water and absorb them. These prevent the inundation of low lying areas. They are also assist the recharging of ground water.

Functions of Wet Lands

Wet land ecosystem mainly depends upon how long it will be under water. The flora mainly receive their nutrients in the form of sediments for their survival. The movement and distribution of the gases, nutrients, detritus, sediments, pollutants and plank are through the medium of water into the system and out of it, the entire biological system depends upon the frequency and duration of flooding physical turbulence, sediment size, deposition and circulation.

Inspite of the scanty availability of nutrients, wet land plants are capable of accumulating the heavy metals and nutrients. The nutrient circulation is very high in wet lands. The microbes are also active in the wet land ecosystem.

The wet lands also receive effluents from industries and flora are capable of absorbing the various heavy metals discharged and use the nutrients make the out flow water free of pollutants.

The wet lands are highly productive areas since they receive nutrients, plants absorb them and support variety of giving out part. Thus energy is trapped and exported. Thus the availability of life supporting elements promote the growth of biological communities. They are considered as nursery grounds of several species of fish and shell fish. Some of the animals spend a part of their life during the migration from the sea (Anandromine) or towards the sea (Catadrornine) in the estuarine region. Variety of organisms occur because this happens to be a transitional zone between land and water.

Effect on Wet Lands by Human Activities

From time immemorial when man realised the importance started converting the wet lands for agriculture, habitation and industrial dumping. The draining and filling leads to the destruction of ecosystem reducing the capacity to receive the flood water, reduction in the ground water recharges and increase in the pollution in adjoining aquatic regimes.

The indiscriminate use of fertilisers and pesticides for agriculture disturbs the wet !and communities killing the non-target species and influencing the primary productivity.

The construction of dams, roads, bridges and bounds upset the movement of water influencing the growth of biota. Leading to eutrophication and depletion of oxygen. It also leads to the breeding of undesirable insect population leading to the transmission of epidemic of diseases.

The dredging operations and discharge of effluents lead to the imbalance in the ecology of wet lands leading to the destruction of ecosystem.

MANAGEMENT STRATEGY

There is a great need to regulate and formulate policies for the management of wet lands. Appropriate agency should be identified to delinate the exact area under wet lands. A thorough study of physico-chemical and biological characteristics including Geology, Hydrology and water quality should be made to monitor the activity in wet lands by the private individuals and Government agencies.

The National Wet land management committee has been constituted and their terms of reference are :

1. To lay down broad policy guidelines for implementing programmes of conservation, management and research of wet lands.
2. To decide priority of wet lands to be taken for intensive conservation measures.
3. To monitor the conservation programmes of management.
4. To prepare Inventory on Indian Wet lands.

CHAPTER 3

Landscape Ecology

Ecological systems do not exist as discrete units but represent different parts of a natural continuum in the form of landscape. Landscape can also be considered as a higher level of biological complexity, immensely useful for understanding various complex processes. Landscape ecology allows the study of these processes on different scales and time. While biodiversity is usually considered at the species level, maintenance of biodiversity requires management at higher levels of organization, particularly at the landscape scale. This is so because management of individual threatened species is difficult. The management should thus focus on the ecosystems that contain these species and on the landscapes in which ecosystems are found. The relatively new discipline of landscape ecology provides insight into both landscape diversity and species diversity and suggests a theoretical and practical basis for conservation of planning.

Forman and Godron (1986) defined landscape as a heterogeneous land area composed of a cluster of interacting ecosystems that is repeated in similar form throughout. Landscapes are composed of a mosaic of patches. Vegetation, a dominant expression of the natural assemblage of various coexisting plant communities, plays a major role in characterizing a landscape with its biodiversity. Species diversity at any one point in a landscape is determined by multiple factors acting on multiple scales (Turner, 1989; Wiens, 1989; Turner and Gardner, 1990). On the landscape scale, the frequency and spatial distribution of critical habitats and resources determine species distribution patterns (Swingland and Greenwood, 1983; Pearson, 1993), while historical accidents, community interactions, and spatio-temporal variability further limit the distribution that is realized at any given time (Wu and Levin, 1997) (Figure 3.1).

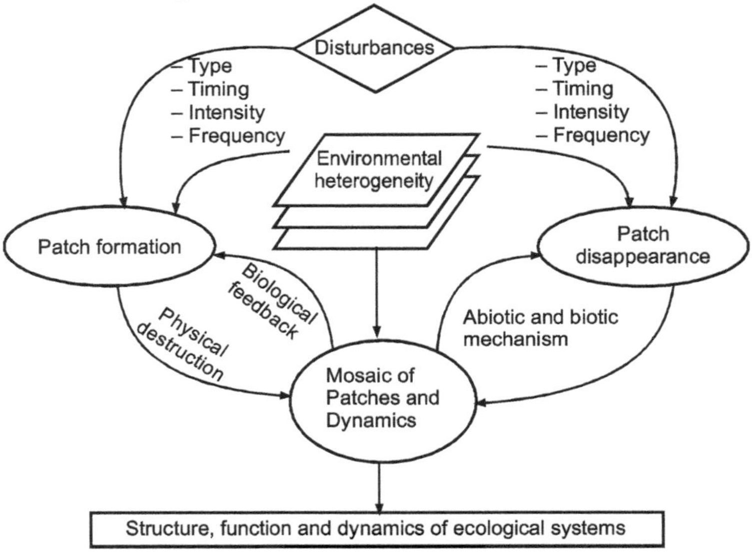

Figure 3.1. Processes and ecological systems in landscape

(Courtesy: SRS & GIS, Eastern Ghats and East Coast, India)

Therefore three basic characteristics of landscapes that affect their diversity: structure, function and dynamics. Structure is the most well understood element of landscapes. It is also the most obvious viewed in the form of different landforms, habitats, or vegetation types. The patch is the basic unit of the landscape structure. The characteristics of patches and the spatial relationships among patches are important components of the landscape (Lidicker, 1995). The distribution of energy, material and species among patches differing in size, shape, abundance and configuration are particularly important to patterns in diversity at the landscape scale. The other two elements of landscapes go beyond description of spatial heterogeneity. *Function* is concerned with interactions among spatial elements of a landscape. *Landscape dynamics* include characteristics of structure and function both in order to examine changes in pattern and processes over time. The conservation and management of biodiversity require understanding of all three elements, including the effects of human activities on the system.

LANDSCAPE DESCRIPTION

At the landscape level, there are two basic components of structure, composition and pattern. Composition refers to the parts (*i.e.*, patch types) that makeup the landscape, and pattern refers to how these patches are arranged. Though these two components are conceptually different, in practice. They are often closely related in space and time (Figure 3.2).

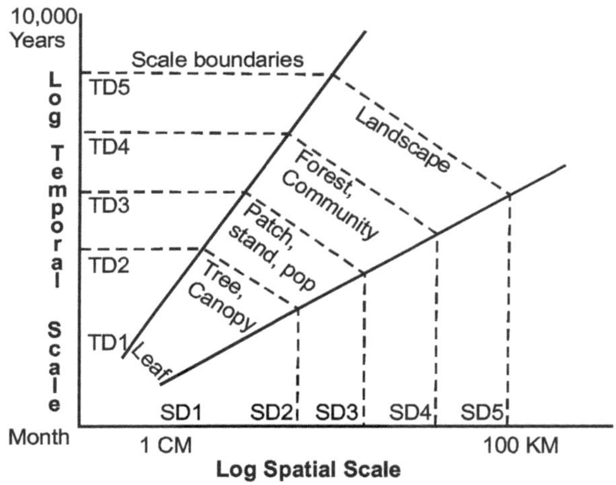

Approximate matching of temporal and spatial scales in ecological studies. Domains of the scales are represented in dotted lines. SDI refers to spatial domain, TDI refers to temporal domain 1

Figure 3.2. Matching of temporal and spatial scales in ecological studies (Adapted from Delcourt et al., 1983)

Landscape composition can be measured in ways analogous to measurements of species composition (Romme, 1982). The most straight forward approach is landscape richness *i.e.*, the number of different patch types in a landscape. Another approach includes the relative abundance or dominance of different patch types along with richness. Measurements of landscape diversity are analogous to common measurements of species diversity (Whittaker, 1960, 1972 & 1977) (Figure 3.3). Different patch types provide different habitats and species composition, thus one might expect that the total number of species in a landscape would increase as landscape richness increases (Burnett et al., 1998).

PATCH DESCRIPTION

A patch is relatively homogenous area that differs from its surroundings. The definition and identification of individual patches and their boundaries are important steps in characterizing the structure of a landscape. In some systems, boundaries may be easily identified, such as patches of agriculture field and adjacent forest in human dominated system, in many cases, however, the boundary is not so clear and patches are more difficult to delineate.

Most methods of patch identification combine qualitative and quantitative approaches. A subjective determination of how different two areas must be, in order for them to be considered different patches is, often needed Turner and Gardner (1990) provided quantitative techniques to group similar cells into homogenous patches or to identify repeating patterns across a landscape. Approaches such as blocking techniques, spectral analysis, and nearest neighbour analysis are often used. These methods include moving window analysis and satellite image to characterize landscapes with sharp transitions.

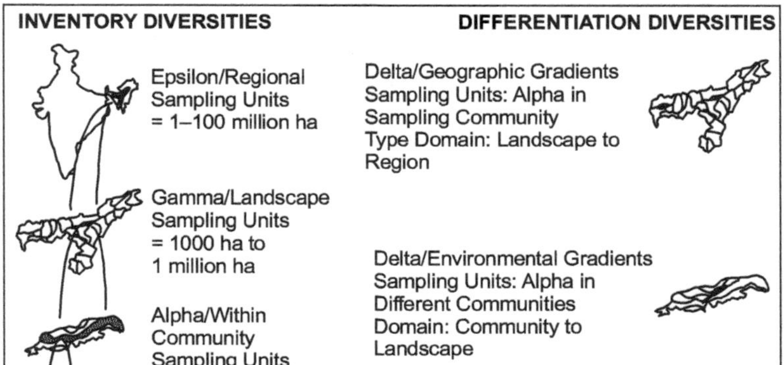

Figure 3.3. Levels of species diversity (Whittaker, 1977)

Once the patches in a landscape have been identified, there are many ways to describe and quantify them (Ritters et al., 1995).

The patch size and shape are the most understood landscape characteristics with species diversity. The relationship between patch size and species richness goes well beyond the familiar species-area curve (Figure 3.4). Although the number of species present in a patch tends to increase with patch size upto a certain limit, the kind of species found also tends to vary with size (Peter and Goslee, 2001). In an important study of tropical deforestation in Amazon forest, species in patches of various sizes were compared to evaluate the importance of patch size to species number (Lovejoy et al., 1984, 1986). Large patches were the richest in species and small patches were found to contain only edge conditions species.

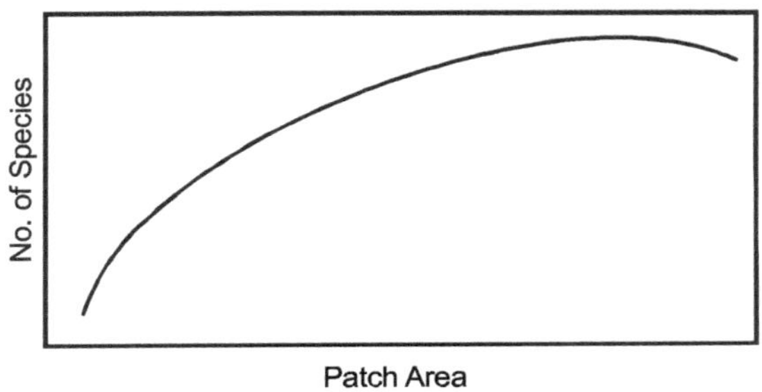

Figure 3.4. Species area curve showing the increase in species number with increasing patch size tending towards a regional limit (Peter and Goslee, 2001)

Patch shape has important effects on different species, including trees, insects, birds and mammals, that were noticeable in a short time. A simple measure of patch shape is perimeter to area ratio. This measure is often standardized so that the most compact possible form, either square or circle is equal to 1. More complex shapes have increasingly higher numbers. Another common index of shape complexity is the fractal dimension, which is also derived from the perimeter and area of a patch. The overall flow of species and resource depends on larger interior or irregular patch shape (Peter and Goslee, 2001).

Fragmentation

The native habitats are converted for human use, (largely as splinters) and the remaining habitats tend to exist as small fragments afloat in a human dominated matrix (Figure 3.5). Hence fragmentation of landscape results in geographical isolation and the probability of recolonisation strongly depend on the distance of fragments from the main core and on the quality of the surrounding habitat.

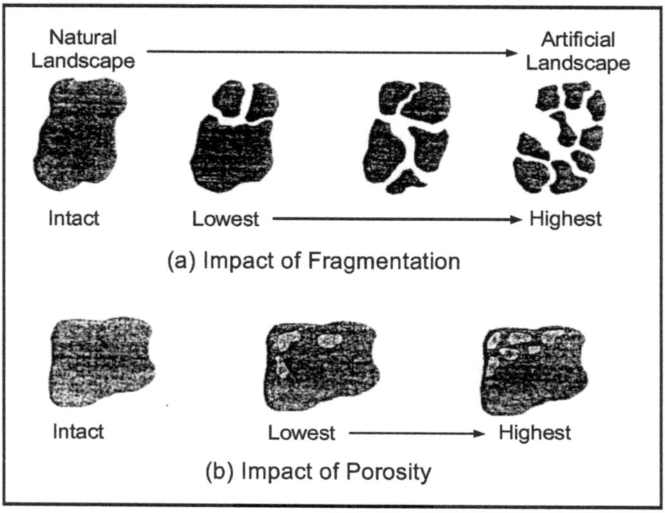

Figure 3.5. Pattern of landscape fragmentation

Fragmentation study takes into account connectivity (corridors), presence of ecotones, the meta population structure etc. Fragmentation increases the vulnerability of these patches to external disturbance with threat on the survival of these patches and on the supporting biodiversity (Nilson and Grelsson, 1995). Fragmentation reduces the size and quality of the habitat over a period, with loss of corridors and continuous impact of edges (Figure 3.6).

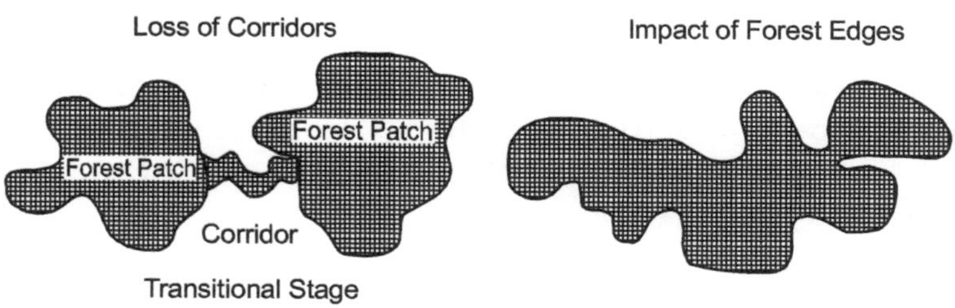

Figure 3.6. Loss of corridors and impact of edges

Large fragments have more species, are less distributed and have lower road access than smaller fragments. Severity of tropical forests fragmentation has been studied with respect to distance effects, fragment size, edge effects and biotic changes (Bierregaard et al., 1992). As a result of fragmentation, there is higher probability of biodiversity loss in tropical regions than temperate regions (Turner and Corlett, 1996).

The corridors are physical and functional connectivity to allow the movement of plant species and fauna. They are severely affected due to fragmentation (Villard and Taylor, 1994; Villard et al., 1995). It is also observed that habitat requirement for sensitive species are specific to area size and surrounding characters (Bancroft et al., 1995). Matrix is the most connected element and structural attribute. Hence, in a landscape perspective, matrix and patches are the elements that are used while considering fragmentation in a landscape (Baudry, 1984; Wiens, 1994). The pathway of loss in biodiversity due to fragmentation in a landscape is shown as Figure 3.7.

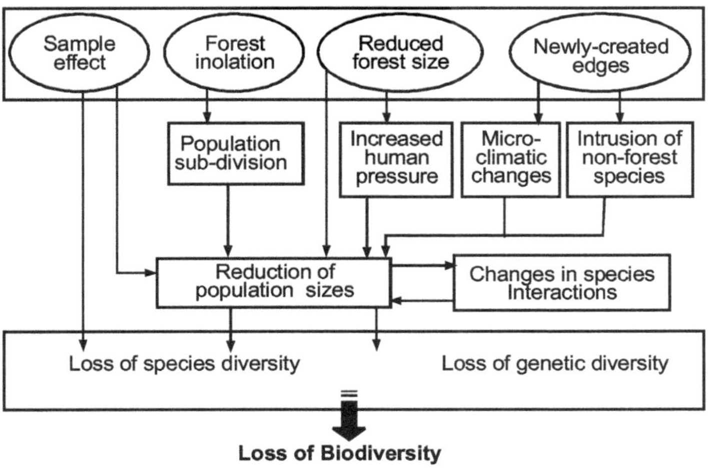

A schematic representation of the theoretical ways in which forest fragmentation may cause reduction in population size may result ultimately in loss of biodiversity (Zuidema et. al., 1996)

Figure 3.7. Impacts of forest fragmentation

Patchiness

Diversity and heterogeneity are two related concepts in landscape ecology. Using the neighbourhood concept, it is possible to measure the relative size and isolation of the patches. Environmental patchiness can reflect a mosaic of soil types, topographic conditions, microclimate and successful stages after recovery from disturbance (Lambeck and Sounders, 1993). The functional relationships of patchiness with species diversity are due to (a) dependence of species diversity on physical substrate, (b) interspecies dependence, (c) species interactions within communities, (d) incidental species relations, and (e) dynamics of a landscape.

DISTURBANCE VIS-A-VIS BIODIVERSITY

Disturbance is a common and wide-spread phenomenon in nature and may be defined as a discrete event along the passage of time that modifies landscape, ecosystems, community and population structure. (White and Pickett, 1985). The disturbance also leads to processes like fragmentation, migration, local and regional extinction. Fragmentation has a strong influence on the dynamics and fate of the material and energy moving across a landscape. The basic variables of disturbance are magnitude, frequency, size and dispersion. Disturbance regimes and their impact on communities and landscape can be well understood by analyzing spatial and temporal architecture of disturbance (Moloney and Levin, 1996). At landscape level, disturbance is related to patch structure and spatial arrangement and determines the fate of patches, their size and duration. Severe disturbance or even a prolonged absence of disturbance generally has adverse effect on biodiversity, but intermediate disturbance seems to enhance diversity in a system (Pickett and White, 1985). Disturbance is the driver of the landscape dynamics and acts at all spatio temporal scales. Human induced disturbance differs from natural disturbance especially in extension, severity and frequency. The capacity of the landscape to incorporate human disturbance is overwhelmed and the disturbance process is transformed into a stress process, which reduces biodiversity. Human activity has widespread impact on biodiversity, affecting ecological entities from species to whole communities and ecosystems, though heterogeneity in the landscape can be due to moderate disturbance. The disturbance regimes can be measured by using different indices i.e., degree of fragmentation, fractal dimension, contagion, juxtaposition, evenness and patchiness (Li and Reynolds, 1994).

BIODIVERSITY CONSERVATION PLANNING AT LANDSCAPE LEVEL

To preserve species diversity most effectively, management plans must preserve the habitats and landscape structure needed by the target species, rather than simply preserving the species in isolation from the larger, potentially changing environment. Management practices aimed directly at a particular species run the risk of losing ecosystem functions which might actually be crucial for the target species, but which were unknown when the management plan was created. Furthermore, maximizing benefits for one species may threaten others. The ideal is to preserve overall ecosystem health, including species diversity. Given the large number of species on the planet, it is impossible, or at best impractical, to manage for every one of them. Instead, conservation biologists are now trying to identify ways to simplify the task of landscape level management.

The primary requirement to evolve such an approach would be:

- To determine disturbance regimes;
- To have knowledge on spatial distribution of biological richness;
- To determine the optimum size of conservation areas; and
- To identify a set of focal species, sensitive to particular landscape structure and function.

CHAPTER 4

Soil Erosion and Soil Depletion

Soil erosion still is are of the worlds major environmental problems. Worldwide loss of top soil amounts to 23 billion tons a year, and is increasing as human population growth forces agriculture to move into marginal, erodible land. Each year, the countries of the former Soviet Union abandon an estimated one-half million hectares of cropland because of soil erosion. Dust storms in North Africa carry upto 100 million tons of soil annually out of the Atlantic, speeding desertification. The Ganges in India and the yellow river in China carry billions of tons of soil to the Ocean, as does the Mississippi river of USA.

Serious soil erosion is forcing agriculture to come up with new techniques. Typically crop production involves plowing land and cultivating crops in raws to control weeds. These expensive, time consuming practices expose to soil to severe erosion. New techniques called conservation tillage systems, range from reduced tillage to no-tillage systems. No tillage systems plant in the residue of the previous crop. An initial application of herbicides eliminates existing grass and weeds and the need for raw cultivation. Conservation tillage significantly reduces soil erosion and the loss of soil nutrients. It increasing the organic matter in soil and encouraged higher populations of soil microbes and invertebrate fauna.

SOIL DEPLETION - SOIL EROSION

Soil is the transitional material between the atmosphere and the earth's surface and the bed rock below the surface. As the material in which plant roots reside, and from which plants extract water and nutrients, soil is essential to the whole life support system on the land. Moreover, soil is the material on which the landscape rests. Although this point seems so apparent that it hardly merits mention, it is nevertheless important to recall that buildings and other structures depend on soil for support and stability. Where supportive strength is weak, as in parts of Mexico city and Veince, Italy, this role of soil is deeply appreciated. Although the formal study of soil is traditionally associated with agronomy, today it is also studied in many other fields, including Civil engineering, Geology and Geography. Agronomist, for example, would limit soil to the solum, the near surface material capable of supporting plant life. The soil contain gaseous, liquid, and solid portions in both inorganic or organic forms, and the Agronomist is interested in what combinations of the materials are best suited for crops. In the Geosciences there are several major objectives in the study of soil. Geologists have a traditional interest in the origin of soil materials as one way of deciphering the geologic history of the land. Civil engineers, by contrast, are concerned about the performance of soil when it is used as footing for buildings or highways. Therefore, Civil engineers define soil on the basis of certain properties that cantor the soils 1) stability when molded into slopes 2) Compressibility under the weight of structures; and 3) Capacity to limit the seepage of water in and around dams, Unlike Agronomists, Civil engineers do not limit their definition of soil to the root zone. Rather, all assemblages of loose materials that over-lie the bedrock are recognized as soil.

What is the physical geographer's objective in the study of soil? Actually, it incorporates parts of all those mentioned above, but three particular areas of interest can be singled out: 1) the role of soil; in the land scape system as it inter-relates with plants, water and atmosphere ; 2) the changes in soil due to erosion and deposition ; and 3) the interplay of soil and land use as in the case of urban development in areas of soft, wet soils.

PHYSICAL CHARACTERISTICS OF SOIL IN GENERAL

The mass of loose material that rests on the bed rock, the stuff in which soil forms, varies greatly in thickness and composition from place to place. In the broad, fattish interiors of the continents, it is usually 25-50m thick and very often several hundred meters thick. In contrast, in areas of rugged terrain, the soil materials may be absent altogether. This is especially so in the high mountains, where erosional processes remove particles of rock from steep slopes as rapidly as they appear on the surface. These particles are moved downslope and accumulate in the valley bottoms, where they often create small areas of extraordinarily deep deposits. Higher on the mountain slopes, barren rock may give way to mixture of snow, ice and rock rubble, all legitimately soil materials.

The solum is the part of the soil capable of supporting plant life. More than half of the volume of most soils is madeup of inorganic and organic particles ; the remainder is mostly air and water.

The ratio of organic to inorganic content decreases sharply with depth, and organic material largely disappears beyond a path of 1-2m in most soils, Dead organic matter is consumed main by microflora near the soil surface the amount of air in a Soil with the water content, and below the water table virtually all end has been displaced by water. Texture is a critical soil property cause of its influence on drainage, moisture holding capacity, a bearing capacity, in particular. The energy that drives the initial soil processes is the heat from solar and geothermal source from the K.E. of moving water, and the molecular energy of chemical ceases. Water enters the soils by infiltration and in the soil is takenup as a capillary water.

SOIL FORCING PROCESSES

Soil Horizon

Working in combination, the chemical, biological, and physical process produce vertical differentiations within the 5 m or so of soils in which there is comparatively free move of gravity water and capillary moisture. This results are in nearly horizontal layers, or horizons, as they are by the soil scientist, which are distinguishable on the basis of colour, texture and chemical composition. Generally speaking, four standard horizons are recognized and they are designated top to bottom by the letters D, A, B, and C. The D horizon includes the humus rich top soil, where the organic content is greater than 20%. The horizon corresponds to the zone of evaluation. The B horizon represents the zone of evaluation and C is transitional to the parent material usually, some sort of deposit, where alteration by soil processes is markedly less advanced. This sequence of horizons, called a soil profile, provides a suitable frame work for describing most soils, particularly those formed in well drained sites in regions of positive moisture balances. As with so many other natural phenomena, the boundaries between soil horizons are typically distinct, as one horizon usually grades almost imperceptibly one into another. These transitional zones are classified as sub horizons, and several are usually identified in each of the upper two horizons (Figure 4.1).

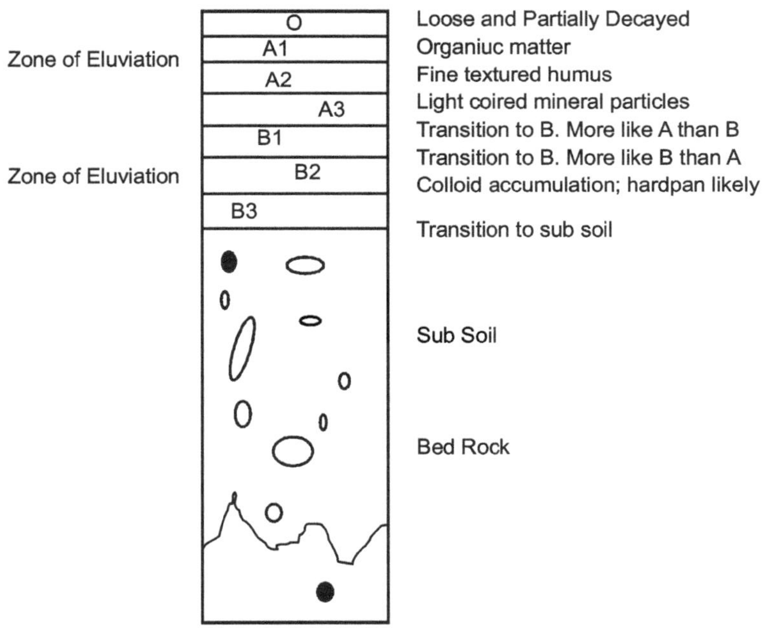

Figure 4.1. A Standard Soil Profile

FACTURE OF SOIL FORMATION

Soil formation is influenced by a multitude of factors, some of which are apparent at a glance, whereas others are very subtle and apparent only through instrumental observation. In all cases, however, these factors can be related to driving forces, which in turn can be translated into basically four types of energy: potential, kinetic, chemical and heat. The processes that operate within the soil, such as eluviations, illuviation, and microflora activity, are driven by this energy, variations in the amounts and forms of energy are controlled by the large environment within which soil formation occurs.

This environment supply energy to soil processes, time is usually also mentioned as a soil forming factor. The following Table 1 provides a summary on the soil forming factors and the energy and processes associated with each.

Table 4.1: Major Factors of Soil formation

Factors	Energy Forms	Soil Processes
Climate	Heat	Erosion, deposition
Parent	Kinetic (water)	Humus formation
Material	Chemical	Microfloral action
Vegetation and related biological activity, chemical	Minerals	Leaching Eluviation
Topography and drainage.	Organic compounds potential (slope)	Illuviation Weathering

SOIL EROSION PRINCIPLES

Two major types of erosion are geological erosion and accelerated erosion. Geological erosion includes soil forming as well as soil eroding processes which maintain the soil in a favourable balance, suitable for, the growth of plants; Accelerated erosion includes the deterioration and loss of soil as a result of man's activities. Though both types of soil removal are recognized, only accelerated erosion is considered in conservation activities.

GEOLOGICAL EROSION

Geological erosion, sometimes referred to as natural normal erosion, is found when the soil is in its natural environment under the protective cover of native vegetation. This type of erosion has contributed to the formation of our soils and their distribution on the surface of the earth. This long time eroding process caused most of our present topographic features, such as canyons, stream channels and valleys.

Practically all geological erosion takes place as a result of the action of water, wind, gravity, and glaciers. Water causes erosion through sheet runoff, stream flow, wave action, and ground water flow. Wind picks up and transports soil particles, thus causing a general mixing of the soil at the surface. Gravity causes mass movement, such as soil creep, rock creep, need flow, rock slide and subsidence of the soil surface.

ACCELERATED EROSION

Accelerated erosion is soil loss in excess of geological erosion. It is normally associated with changes in natural cover or soil conditions and is caused primarily by water and wind. The forces involved in accelerated erosion are 1) attacking forces which remove and transport the soil particles and 2) resisting forces which restored erosion. Hereafter, accelerated erosin will be referred to as soil erosion or simply erosion. Erosion is one of the most important agricultural problems in the world. In addition to soil losses there are large losses of plant food, elements which are removed in the runoff.

WATER EROSION

Water erosion is the removal of soil from the land's surface by running water, including runoff from melted and ice water erosion subdivided into raindrop, shut, rill gully and stream channel erosional.

FACTORS AFFECTING EROSION BY WATER

The effect of precipitation on soil and water losses is illustrated in Figure 4.2. There is a direct relationship between total runoff and soil loss from agricultural areas, the factors affecting runoff also influence vegetation and topography. Of these, the vegetation and to some extent the soil may be controlled. The climatic factors and the topographic factors, except slope length, are beyond the power of man to control.

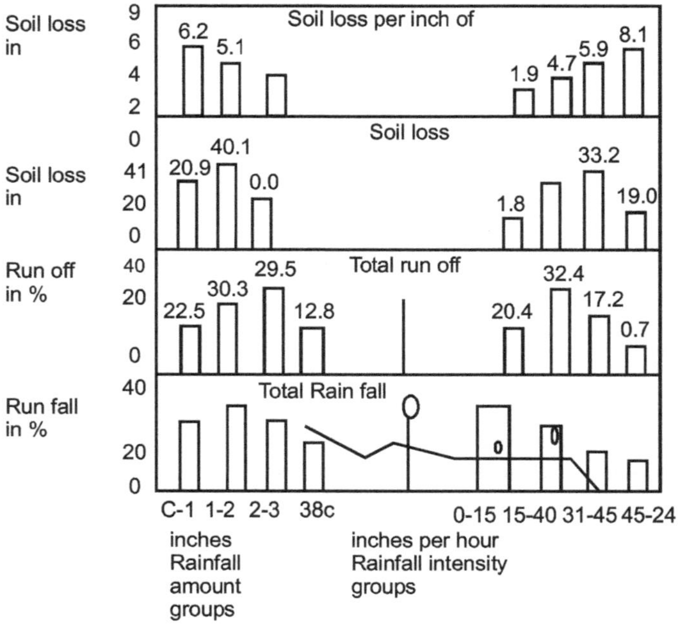

Figure 4.2: Effect of Rainfall and Runoff on Soil Loss from a Sandy Clay Loan Soil in North Carolina, USA.

CLIMATE

Climate factors affecting erosion are precipitation, temperature, wind, humidity, and solar radiation. Temperature and wind are most evident through their effects on evaporation and transpiration. However, wind also changes raindrop velocities and the angle of impact. Humidity and solar radiation are some what loss directly involved in that they are associated with temperature.

Soil loss and runoff can be compared for different amounts and intensities of rain. Total soil loss varies with the total runoff for different rainfall amount groups. The soil loss per inch of runoff, however, decreases as the size of the storms increases. When storms are grouped by rainfall intensity, the soil less ear inch of runoff increases as the intensity increases, even though the total rainfall decreases. In many instances one or two high inensity rains may cause as much soil loss as all other storms during a season of during several seasons.

SOIL

Physical properties of soil affect infiltration capacity end the extent to which it can be dispersed and transported. These properties which influence erosion include soil structure, texture, organic matter, moisture content, and density or compactness, as well as chemical and biological characteristics of the soil. As yet, no one soil characteristic or index provides a satisfactory means of predicting erodibility.

VEGETATION

The major effects of vegetation in reducing erosion are 1) interception of rainfall by absorbing the energy of the raindrops and thus reducing runoff 2) retardation of erosion by decreased surface Velocity, 3) physical restraint of soil movement 4) improvement of aggregation and porosity of the soil by roots and plant residue; 5) increased biological activity in the soil, and 6) transpiration, which decreases soil moisture, resulting in increased storage capacity. These vegetative influence vary with the season crops, degree of maturity, soil and climate, as well as with the kind of vegetative material, namely, roots, plant tops, and plant residues.

TOPOGRAPHY

Topographic features that influence erosion are degree of slope, length of slope, and shape of the watershed. One steep slopes high velocities cause serious erosion by scour and by sediment transportation.

RAINDROP EROSION

Raindrop erosion is oil splash resulting from the impact of water drops directly on soil particles or on thin water surfaces. Although the impact on water in shallow streams may not splash soil, it does cause turbulence, providing a greater sediment carrying capacity.

Tremendous quantities of soil are splashed into the air, most of it more than once. The amount of soil splashed into the air as indicated by the splash losses from soil elevated pans was found to be 50 to 90 times greater than the washoff losses. On bare soil it is estimated that as much as 100 tons of soil per acre are splashed into the air by heavy rains.

Characteristics of raindrops affecting erosion are drop size and velocity. Large drops may increase the sediment carrying capacity of runoff as much as 12 times. The velocity of raindrops greatly affects soil splash and erosion. From a study of fair soils, it is found that the relative soil loss for a 30 minutie rainfall period varies as the 4.33 power of the velocity.

The process of soil erosion involves soil detachment and soil transportation. The corresponding soil characteristics that described the case with which the soil particles may be detached and transported are soil detachability and soil transportability. In general, soil detachability increases as the size of the particles increase, and soil transportability increases with a decrease in particle size. That is, clay particles are more difficult to detach than sand but clay is more easily transported.

SHEET EROSION

Sheet erosion is the uniform removal of soil in thin layers from sloping land. This results from sheet or overland flow, the runoff from the surface in thin layers. Although important, sheet erosion is of ten unnotices because it occurs gradually. The beating action of rain drops combined with surface flow causes the major portion of sheet erosion. From an energy stand point, rain drop erosion is for more important because raindrops have velocities of about 20 to 39 fps. Whereas overland flow velocities are about 1 to 2 fps. Raindrops cause the soil particies to be detached and the increased sediment reduces the infiltration rate by scaling the soil pores. Areas where loose, shallow top soil over lies a tight sub soil are most susceptible to sheet erosion. The eroding and transporting power of sheet flow are functions for the depth and velocity of runoff for a given size, shape, and density of soil particle or aggregate.

RILL EROSION

Rill erosion is the removal of soil by water from small but well defined channels or streamlets where there is a concentration of overland flow. There is no sharp line of demarcation where sheet erosion ends and rill erosion begins. Rills are small enough to be easily removed by normal tillage operations. Detachability and transportability are both greater in rill erosion than in sheet erosion because of high velocities. Rill erosion is most serious where intense storms occur on soils having high runoff producing characteristics and loose, shallow top soil.

GULLY EROSION

Gully erosion produces channels larger than rills. These channels carry water during and immediately after rains and as distinguished from, rills, gullies cannot be obliterated by tillage. Thus, gully erosion is an advanced stage of rill erosion much as rill erosion is an advanced stage of sheet erosion.

PRINCIPLES OF GULLY EROSION

The rate of gully erosion depends primarily on the runoff-producing characteristics of the water shed; the drainage area, soil characteristics, the alignment, size, and shape of the gully, and the slope in the channel.

A gully develops by processes that may take place either simultaneously or during different periods of its growth. These processes are 1) Water fall erosion at the gully head, 2) Channel erosion caused by water flowing through the gully or by raindrop splash on unprotected soil, 3) alternate freezing and thawing of the exposed soil banks, and 4) slides or mass movement of soil in the gully.

CLASSIFICATION OF GULLIES

Of the several systems of gully classification, the one given in Table 2 is based on an arbitrary classification gully sizes and drainage areas as small, medium, and large.

Table 4.2: Description of Gullies and Drainage areas.

Description	*Gully depth (feet)*	*Drainage area across*
Small	3 or less	5 or less
Medium	3 to 15	5 to 50
Large	15 or more	50 or more

Another system classifies gullies with respect to their cross sections. Gully cross sections may be V-or U - shaped, depending upon soil and climatic conditions, age of the gully and type of erosion. U-shaped gullies may be found in loose soil regions and alluvial valleys where both the surface soil and the subsoil area easily eroded. Under such conditions, gullies tend to develop vertical walls which result from undermining and collapse of the banks. The scouring of soil by concentrated runoff unprotected/ depressions results in V-shaped gullies having sloping heads. Such gullies may develop where the subsoil is resistant to erosion. Both V and U-shapes are commonly found in the same channel.

Stream channel erosion consists of soil removal from stream banks or soil movement in the channel. Stream channel erosion and gully are distinguished primarily in that channel erosion applying to the lower end of head water tributaries and to steams that have nearly continuous flow and relatively flat gradients where a gully erosion generally occurs in intermittent streams near the upper ends of head water tributaries.

WIND EROSION

Wind erosion, unlike water erosion cannot be divided into types, as wind erosion varies only by degree. For example there may be only a slight disturbance or the surface coexerting a small area of there may be a huge dust storm covering several states. Wind erosion takes place at slow, geologic rate under actual vegetation and normal soil conditions, whereas under cultivation serious erosion may result.

FACTORS AFFECTING WIND EROSION

The major factors affecting wind erosion climate, soil, and vegetation. Topography appears to be relatively unimportant in the wind erosion process, but the length of the eroding surface greatly influences soil movement.

CLIMATE

The climatic factors influencing wind erosion are precipitation, temperature, wind; humidity, viscosity and density of the air. The amount and distribution of rainfall and its effect on soil moisture are of primary concern. The dispersing and mellowing action of freezing and the wind may be of considerable importance, especially in the spring. Evaporation and transpiration as influenced by temperature, wind, and humidity, deplete soil moisture, thus making the soil more subject to wind erosion. The principal characteristics of wind affecting erosion are velocity, direction, duration and turbulence.

SOIL

The soil factors affecting wind erosion are texture structure, density of soil particles, density of soil mass, organic matter, moisture content, and surface roughness. The moisture content is especially significant because only a relatively dry soil is subject to wind erosion. Surface roughness changes with tillage practices such as listing and plowing. Surface crusts, although decreasing surface roughness have a retarding influence on soil movement.

VEGETATION

The vegetative factors are the height and density of cover, type of vegetation and seasonal distribution. Living plant roots and tops are more effective in retarding erosion than crop residues.

TYPES OF SOIL MOVEMENT

Suspension, saltation (sediment movement by saltation occur's where particles skip or bounce along the stream bed), and surface creep in wind erosion are comparable to suspension, saltation and bed load, respectively in sediment movement by water.

These three types of movement usually occur simultaneously. Investigations have shown that the major portion of soil movement take place near the surface at height not greater than approximately 8 feet. Above this height the only movement is normally by suspension, whereas all the three types of movement occur near the surface.

Movement of soil particles by saltation is caused by the pressure of the wind on the soil particle and its collision with other articles. The characteristic path of a soil particle moved by saltation together with final and initial velocity, vectors are shown in Figure 4.3. As the soil particles leaves the surface, it moves nearly in a vertical direction. The horizontal distance through which the particle continues to rise is about one-fifth to one fourth of distance L. As the particle descends to the surface, it travels in a straight line with an angle of descent of about 6 to 12 degrees.

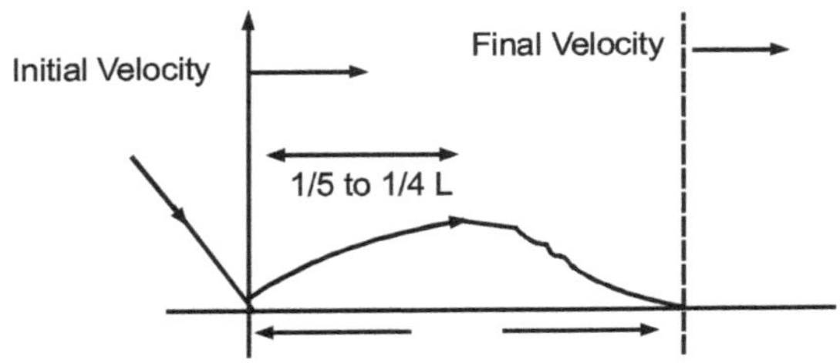

Figure 4.3: Characteristic path of a soil particle moved in air by saltation

MECHANISM OF WIND EROSION

Wind erosion takes place in the three distinct phases, initiation of movement, transportation and deposition. Initiation movement. The soil movement is initiated as a result of turbulence and velocity of wind. The fluid threshold velocity is defined as the minimum velocity required to produce soil movement by direction of the wind, whereas the impact threshold velocity is the minimum velocity required to initiate movement from the impact of soil particles carried in slatation except very near the surface and at very low velocities (less than about 2 mph), the surface wind is always turbulent.

TRANSPORTATION

The quantity of soil moved is influenced by a particle size, gradation of particles, wind velocity and distance across the eroding area. Winds being quite, variable in velocity and direction, produce gusts with eddies and cross currents which lift and transport soil. The quantity of soil moved varies as the cube of the excess wind velocity over and above the constant threshold velocity, directly as the square root of the particle diameter and increases with the gradation of the soil. The atmosphere has a tremendous capacity for transporting soil particularly those soil fractions less than 0.1 mm in diameter. It is estimated that the potential carrying capacity for 1 cubic mile of the atmosphere is upto 126,000 tons of soil, depending upon the wind velocity.

DEPOSITION

Deposition of sediment occurs when the gravitational force is greater than the forces holding the particles in the air. This generally occurs when there is a decrease in wind velocity.

SOIL LOSSES

Soil losses vary considerably with the type of erosion. Often both wind and water erosion are present in the same area. The importance of soil losses is indicated by the effect of top soil. Losses is indicated by the effect of top soil depth on crop yield, as shown in Figure 4.4. Since the data has been compiled from many areas, the average curve and the limit variation do not represent any one condition. At greater depths of top soil, yields are reduced at a slower rate than at the shallower depths. For example at top soil depths of 14, 7 and 0 in cms are 100, 70 and 25%, respectively. In other words, yields were reduced 45% for the lower 7 inches and only 30% for the top 7 inches. On some soils these yields decreases can be largely overcome by proper fertilization.

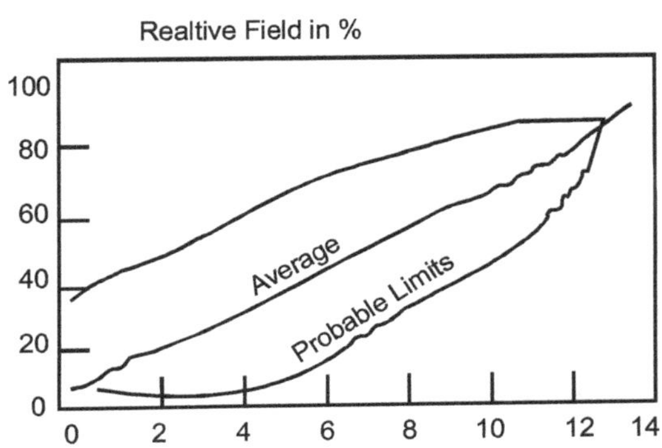

Figure 4.4. Effect of top soil depth on relative crop field

ESTIMATING SOIL LOSSES

It is desirable to enable to predict soil losses for a given set of conditions in order to determine the adequacy of conservation measures in form for planning. Any method for estimating soil loss should show the relationship of each of the factors and should be easily applied to field conditions. The annual soil loss by water erosion may be expressed as function of equation in which

$$X_a = f(K,S,L,R,F,E,C,P) \text{ -I}$$

Where X_a = annual soil loss in tons per acre, K = soil factor, S = Degree of slope factor L = Length of slope factor, R = Rotation factor, F = Fertility factor, E = Degree of erosion factor, C = Conservation of factor, and P = Rainfall factor.

In any area where the amount and intensity, of rainfall is merely uniform, P may be omitted *as* a variable. By evaluating the remaining factors from plot and watershed studies, curves may be developed for each factor as in Figsures 4.5 and 4.6. By using the numerical value for each factor taken from such curves, the soil loss may be computed by the equation

$$Xa = 10\,(KSLRFEC) = 11$$

By properly selection rotations, fertility practices and soil losses can be reduced to the permissible value or less. The loss of plant nutrients from the soil is atleast as important as the loss of the soil itself. The effect of wind as loss of plant nutrients is likely to be many times greater during the first few years after cultivation than later when the fines have been removed. In this way wind erosion may be serious even when the total soil loss is not in itself excessive.

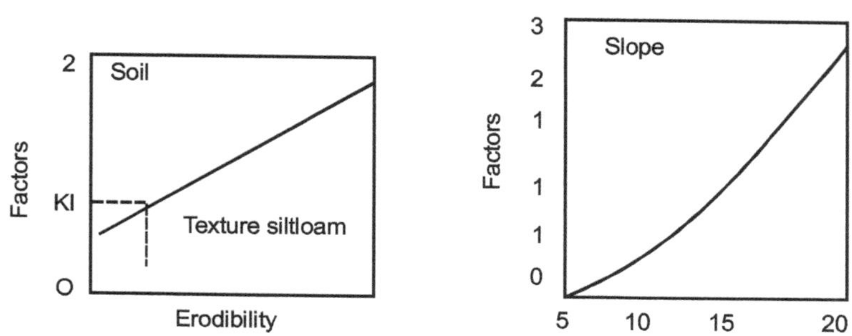

(a) Soil series and types

(b) Slope in %

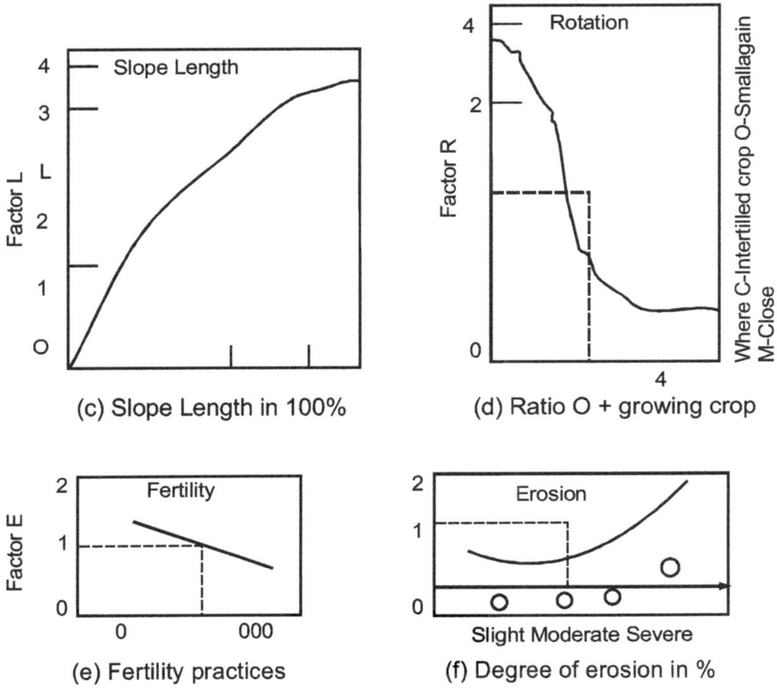

Figure 4.5: Soil Erosion Factors for Soil Type, Slope, Slope Length, Rotation, Fertility and Degree of Erosion

Accelerated erosion, disappearing beaches, increased frequency of flooding progressive siltation, degraded ecosystems and so on are all symptoms of man's inability to land management. However to add to the difficulties, many of the problems may occur naturally.

The loss of coastal land is a pernicious and often perplexing problem. Loss of land has both direct and indirect consequences for coastal communities requiring abound ownment and adjustment cases posing posing a danger of property and in extreme a to life may be ascribed to a variety including sea tectonic instability and subsidene, climatic and numerous man made causes. The later mainly with changes in the sediment badget through dreging, aggregate extraction infilling or impedance of supply.

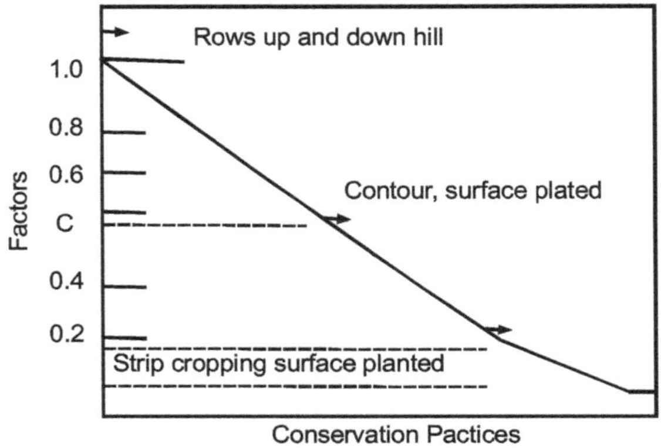

Figure 4.6: Effect of Conservation Pactices on Soil Erosion

METHODS OF EROSION CONTROL ON COASTAL CLIFFS AND SLOPES

By tradition, Engineers relied on interposing static structures between the sea and the shore. Protection against direct storm wave attack requires the construction of a massive barrier. A vertical wall capable of with standing extreme dynamic pressures must be constructed of durable, immovable materials. The sea wall at St. Malo in northern France is an example of this type of wall, it is constructed of interlocking granite blocks, capable of withstanding 6m, breaking waves, although wooden pilings have been inserted on the beach in front of the wall in an attempt to pre-break the waves and so reduce the dynamic on the wall itself. Most sea walls to be secured by pilings while some require additional too protection against undercutting by waves. Back shore protection is also sought by constructing rubble mound barri cades. Tetrapods are used widley to protect the shore. Their large irregular surface area helps wave dissipation while 'legs' interlock to produce stability.

Some environmental problems related to sea walls are (a) falling beach levels, poor and impeded sediment exchange lead to both undercutting and over stumping (b) at the end of the wall flaking risk of storm damage.

FURTHER READING

1. Richard K. Frevert, Glenn, O. Schwach. Talcott W. and Kenneth K. Barnes (1963). Soil and Water Conservation Engineering. John Wiley & Sons. Inc., New York.
2. William M. Marsh and Jeff Deziern (1981). Landscape, introduction to Physical Geography, Addition-Wesley Publishing Company, Reading, Nwe York.

CHAPTER 5

Desertification and Its Impact on Soils in India

INTRODUCTION

Desert is a land with poor soil fertility. The poor soil fertility is due to scanty rainfall, adverse geo-chemicals or even manmade mistakes. Thus desert can be created or exponded by man, unintentionally. The natural ecological processes in the fringes of a desert, unfortunately promote the expansion of deserts. Thus all deserts, like all other human evils, have a lot of self expansive tendency, if left unchecked. Delays always raise the cost of control efforts. In India, expansion of Rajasthan deserts have been alarming. The food causes a lot of damage to the standing crops and there can be a drought even after heavy rains.

In India, the creation or expansion of deserts from the following factors: Over grazing of land due to fodder scarcity removes the vegetation cover, making the land susceptible to the erosion of the rich top soil. Open mining also has the same effect. Deforestation sets in a chain of reaction of ecological destruction. Desertification is the result of an ecological collapse, under severe ecological stresses. Over pumping of ground water for municipal, industrial and agricultural consumption, lowers the ground water table. In the near future, if the present pollution trends continue, acid rains can destroy the terrestrial an aquatic ecosystem. The antidotes to desertification are obviously aforestation, fodder cultivation, weather management and pollution control.

Desertification is the degradation of dry lands. It involves the loss of biological or economic productivity and complexity in croplands, pastures, and woodlands. It is mainly due to climate variability and unsustainable human activities. The most commonly cited forms of unsustainable land use are over cultivation, overgrazing, deforestation and poor irrigation practices. Seventy per cent of the world's dry lands (excluding hyper-arid deserts), or some 3,600 million hectares, are degraded, while drought is often associated with land degradation. It is a natural phenomenon that occurs when rainfall is significantly below normal recorded levels for a long time. Dry lands respond quickly to climatic fluctuations.

About the billion people live in arid or semiarid desert lands that occupy about one third of the world's land surface. In these drier parts of the world, deserts are increasing rapidly from a combination of natural processes and human activities, a process known as desertification or land degradation.

Over the ages, dry land ecology has become attained to this variability in moisture; plants and animals can respond to it rapidly. For example, Satellite imaginary has shown that the vegetation boundary south of the Sahara can move upto 200 km when a year is followed by a dry one, and vice versa.

The biological and economic resources of drylands, notably soil, quality, freshwater supplies, vegetation, and crops are easily damaged. People have learned to protect these resources with age-old strategies such as shifting agriculture and nomadic herding. However, in recent decades these strategies have become less practical due to changing economic and political circumstances, population growth, and a trend towards more settled communities. When land manages cannot or do not respond flexibly to climate variations, desertification is the result.

In many cases, unregulated access to land resources may lead some individuals to maximize their own gains by overexploiting the land at the expense of access to the best land, depending upon the most fragile areas and resources. Their poverty may give them little alternative but to extract what they can from the scarce resources available to them, even though this degrader the land. International economic forces can encourage people to overexploit their land. International trade patterns can lead to the short-tern exploitation of local restoring the land. Similarly, the development of an economy based on cash crops, or the imposition of taxes, can distort local markets and promote overexploitation of the land. Ignorance, errors, and natural and man-made disasters can also contribute to land degradation.

Exposed and eroded topsoil can be blown away by the wind or washed away by rainstorms. The soil's physical structure and biochemical composition can change for the worse. Gullies and cracks may appear and vital nutrients can be removed by wind or water. If the water table rises due to inadequate drainage and poor irrigation practices, the soil can become waterlogged, and salts may build up. When soil is trampled and compacted by cattle, it can lose its ability to support plant growth and to hold moisture, resulting in increased evaporation and surface run-off. Vegetation becomes damaged. The loss of vegetation cover is both a consequence and a cause of land degradation. Loose soil can sandblast plants, bury them, or leave their roots dangerously exposed. When pastures are overgrazed by too many animals, or by inappropriate types, edible plant species may be lost, allowing inedible species to invade. Some of the consequences are borne by people living outside, immediately affected area. Degraded land may cause downstream flooding, reduced water quality, sedimentation in rivers and lakes, and siltation of reservoirs and navigation channels. It can also cause storms and air pollution, resulting in damaged machinery, reduced visibility, unwanted sediment deposits, and mental stress. Wind-blown dust can also worsen health problems, including eye infections, respiratory illnesses, and allergies.

A nutritionally adequate diet for the world's growing population implies tripling food production over the next 50 years. This will be difficult to achieve even under favourable circumstances. If desertification is not stopped and reversed, food yields in many affected areas will decline. Malnutrition, starvation, and ultimately famine may result. The relationship between soil degradation and crop yields, however, is seldom straight forward. Productivity is affected by many different factors, such as the weather, disease and pests, farming methods, and external markets and other economic forces. Desertification contributes to famine. Famine typically occurs in areas that also suffer from poverty, civil unrest or war. Drought and land degradation often help to trigger a crisis, which is then made worse by poor food distribution and the inability to buy what is available. Desertification has enormous social costs. There is how increased awareness of the relationship between desertification, movements of people and conflicts.

OBSERVATIONS

An annual rainfall of less than 25 centimeters will produce a desert anywhere in the world. In the semiarid areas are now unreliable and the lands that were once semiarid are now becoming desert. The process of desertification is precipitated by prolonged droughts, causing the top layer of the soil to dry out and blow away. The eroded soils become unstable and compacted and do not readily allow for seeding. This means that desertified areas not to regenerate by themselves but remain bare and continue to erode.

Natural desertification is greatly accelerated by human activities that leave soils vulnerable to erosion by wind and water. The drier grasslands with too little rain to support cultivated crops have traditionally been used for grazing livestock. When semiarid land is overgrazed (by keeping too many animals on too little hand), plants that could survive moderate grazing are uprooted and destroyed altogether. Since plant roots no longer bind in soil together, the exposed soil dries out and is blown away as dust. The destruction and removal of the topsoil means that soil productivity drops drastically. The obvious solution of desertification caused by overgrazing is to limit grazing to what the land can sustain, a concept that is easy to expose but difficult to practice.

Other human factors leading to desertification include over-cultivation, deforestation, salting of the soil through, irrigation and the plowing of marginal land. These are due to the destructive practices are intensified in India by rapid pollution growth, high population density, poverty and poor land management.

The process of erosion of and soil formation vary with climate and with the composition of the parent material. In balanced ecosystems, soil lost to erosion is replaced by new soil created by natural processes. On overage new soil is formed at a rate of about 5 tonnes per acre (12.5 tonnes per hectare) per year, which is equivalent to layer of soil about 1/64 inch (0.4 cm) thick. This means that soils can sustain an erosion rate upto 5 tonnes per acre per year and still remain in balance.

Forests are cut down for many different reasons. Some are cleared for agriculture, some for construction, some for paper products, some for firewood and heating purposes (Tables 5.1 & 5.2). Unfortunately deforestation results in more than just the loss of trees (Table 5.1) for soil is erosed, nutrients are lost to the ecosystem and the water cycle is disrupted. Based on DCPPAT, 1993, requirements of different categories shown in Table 5.2. The roots of the trees serve to bind the soil together and to hold water in an ecosystem, while the leaves of the trees break the force of the rain and allow it to soak into the topsoil. The result is that surface runoff from a forested hillside is half as much as from a grass covered slope. Additionally, water soaking into the ground (rather than running off) leads to the natural recharge of ground water and other water resources. Water and soil runoff from deforested hillsides cause flooding and siltation of agricultural and aquatic ecosystems in adjacent lowlands. Forests are also more efficient at reabsorbing and recycling the nutrients released from decaying detritus grass are lands. Clearing forests therefore exposes the soil to both erosion and nutrient loss and alters the recharge of water reserves in the ecosystem.

Table 5.1. Uses of Wood in the World (Based on FAO, 1991)

S. No.	*Region*	*Total wood consumption (Billion m^3)*	*%wood use*	
			Industrial	*Firewood*
1.	Global	32	46	54
2.	Developing Countries	1.8(57%)	18	82
3.	Developed Countries	1.4 (43%)	84	16

In developing countries including India, on the other hand, high population growth rates and widespread poverty put pressure on the forests. Trees are needed for firewood deforestation exceeds replanting by five times. For example in Brazil, soil erosion and desertification have resulted from the conversion of forests to cattle ranches. In China about one-third of its agricultural land has been lost to erosion and the story is similar for many other countries.

Planting trees on hillsides is particular effective. Reforesting desertified areas first requires mulching the soil to hold moisture and the protection of the seedlings for several years until natural process can regenerate the soil. Using these methods, Israel has achieved spectacular success in bringing desertlands (a product of past desertification) back to agriculture.

Causes

Causes of desertification are (1) climate factors, (2) human factors (human cultures). Recent changes in land use and population density had much ecological effects. Human factors are population growth and increased density, reduced nomadism and loss of grazing lands, (3) interactions between climate and culture.

The causes of deforestation and denudation are well known. The principle causes have been the population explosion in man and livestock leading to enhanced requirement of timber and fuelwood (Table 2) and grazing respectively. Increasing number of consequent devastation. The most accessible forests areas are heavily grazed. For instance there are nearly 12,00,000 sheeps and goats in alpine areas of Uttar Pradesh state and in addition there are also visit about 25,000 migratory graziers. There are also about 57,000 buffaloes owned by Guljaras. It is estimated that in absence of suitable checks, growing forestry stock will decrease from 13.79m^3/head in 1981 to 2.66 m^3 head in 2001; and reduction in annual availability of gross will be from 2.60/t/cattle unit in 1981 to 0.90/t/cattle unit in 2001. Existing forests may thus not be able to meet our wood and fodder requirements of mankind and livestock respectively.

Table 5.2: Requirement of wood (mt) by Man (Based on DCPPAT, 1993)

S.No.	*Category*	*Requirement in 1980*		*Requirement in 2000*		*Additional Requirement*	
		m^3	*% of Total*	*m^3*	*% of Total*	*m^3*	*% of Total*
1.	Fuel Wood	188.600	87.5	225.000	78	36.4	19
2.	Timber	22.720	10.5	46.755	16	24.0	106
3.	Pulp wood	4.175	2.0	17.695	6	13.5	323
	Total	215.495	100	289.450	100	73.9	34

Moreover, increase in podu or shifting (Jhum) cultivation in Andhra Pradesh and Orissa state has also laid large forest tracts lost. In essence, it is a low energy budget and low investment system of subsistence agriculture that uses rapid mineralization and recycling of nutrients. There has been shortening of podu or shifting (Jhum) cultivation cycle to six years only (in some districts, even 2-3 years only), which provides no enough time for natural repair of damaged ecosystem, already spoiled.

Forests have since long been looked upon as revenue generation sector and maximum could be extracted from the trees by Government and private owners. In the face of agriculturalisation, urbanization and industrialization preservation of forests could be given a very low priority. India lost during 1951 to 1956 about 4.2 million hectares of forests for such activities. Agriculture claimed maximum of about 2.5 million hectares.

A major cause of deforestation in India has been the construction of hill roads about 30,000 km long, most of which are in strategic and most fragile belt of Himalayas. Road construction affected the stability of hill slopes, damaged the protective vegetation, orchards and agriculture fields, blocked natural drainage, polluted steams, delayed vehicular traffic and caused damage to human life and property.

When drought occurs, dust is created leading to desertification, because its dust particles lead to dessication and dry the land. This is a comman phenomenon in India. One hectare of tree plants reduces the 50 tonnes of dust.

RESULTS & DISCUSSION

Desertification process is leading to desert formation. This may result either due to a natural phenomenon linked to climatic change or due to abusive land use. In fact even for climate change, these are the improper land use practices which are largely responsible. Removal of vegetation cover brings about marked changes in the local climate of the area. Thus deforestation, overgrazing etc., bring about changes in rainfall, temperature, wind velocity etc., and also lead to soil erosion. Such changes than lead to desertification of the area. Desertification often starts as patchy destruction of productive land. Increased dust particles in atmosphere lead to desertification and drought in margins of the zones that are not humid. Even the humid zones are in danger of getting progressively drier of droughts continue to occur over a series of years. Indications are clear that the temporary phenomena of meteorological drought in India is tending to become permanent one. This tend is not restricted to the fringes of exciting deserts only. The threat of desertification is thus real, because as the forest diminishes, there is steady raise in atmosphere temperature. At the time of independence (15 August, 1947) in the country, 75 million hectares, about 32% was under forest cover. Today this has been reduced to just over 15%. India has been losing 10 million trees every 24 hours. Thus deforestation is one of the main factors leading to desertification, primarily through its effect on climate of the area. As a result of gross mismanagement of natural resources including land, certain irreversible changes have triggered the breakdown of nutrient cycles and microclimatic equilibrium in the soil indicating the onest of desertic conditions.

Acknowledgements

Thanks are due to the Organizers of Consoil, 2003 who accepted paper on "DESERTIFICATION AND ITS IMPACT ON SOILS IN INDIA" as poster presentation (Special Themes on Soil and Ground Water) at Gent, Belgium on 12 May, 2003, presented by the author, Prof. M.V. Subba Rao.

CHAPTER 6

Environmental Modelling

Man is gifted by nature with assets like air, water, land, forests, minerals, fossil fuels, animals and biodiversity. Immense beauty of nature gives him aesthetic pleasures. Man with his intelligence and scientific advancement started exploiting natural resources irresponsibly while releasing pollutants into the environment. Nature could assimilate the pollution and could give resources for thousands of years. The recent pace of advancements of mankind started threatening nature's capacity to meet his needs and greeds. Now environmental degradation and its impact on man's well being has become a global concern. Sustained development, ecological stability, resource conservation, harmony with nature and preservation of our globe drops can be studied in laboratory model. Impact of air pollution on plants can be studied in closed clambers in a laboratory. The impact of a fertilizer can be monitored on a selected plot of the field. Physical models are difficult to duplicate the nature conditions. They are costly. Usually physical models cannot simulate the open system with complex environmental interactions. Physical models are too case specific.

2. Empirical Models: Environmental data collected from the field is processed through regressions and curve fitting to obtain functional relationships among the relevant parameters. Such relationships may not even have physical basis for comprehension and interconnection. For example, the average life span of an animal is five times the time taken to attain mature body size. Such empirical models are only thumb rules for convenience and are useful to serve during management and to stimulate theoretical pursuits in certain directions.

3. Statistical Models: While empirical models are based on limited experimental data, statistical models are based on large volumes of data. Larger the sample size greater is the reliability of a statistical model. The simplest example of a statistical model is that heads will turnup in 50 per cent of coin tossings. Another example is that the percentage of a population with a particular height follows a mathematical equation, called the normal distribution curve. Statistical models are not applicable to particular cases. Only probabilities and trends can be predicted.

4. Mathematical or Analytical Models: These are relationships among parameters, that can be derived from known scientific principles. For example, the weight of a person tends to vary with the cube of his height. Another example is the relation between carbondioxide absorbed and starch produced in plant photosynthesis. Analytical models are in the form of algebraic equations. By substituting the known values of environmental parameters in the equation, the unknown values can be calculated. In environmental modelling, analytical models are very few since the interacting parameters are too numerous and the interaction kinetics is too complicated.

5. Analogue Models: In nature, different phenomena may exhibit similar functional relationships. For example the rate of flow of water in a channel is related to the drop in the level of bed. The rate of flow of electrical charge through a conductor is related to the drop in voltage. Since the functional relationships are similar water flow in a system of channels can be modelled with electrical circuits which are cheaper and changes. Reclamation will severally effect the ecology as habitals are both destroyed and created.

COASTAL HAZARDS

Many coastal areas are at risk from natural hazards, created by geological or meteorological disturbances of the sea surfaces. These risks are of two kinds (a) short term (minutes, hours or days) risks associates with storms or earthquakes (b) long term risks related to changes in rise in level caused by regular eustatic (atmospheric, or isostatic tectonic) factors.

Management of Storm Hazards

This have been many instances throughout the world that cyclones effected property, people, etc. Two recent instances one in Bangladesh in 1970 and second in U.S.A. in 1969 are most serious. The effects may be subdivided into direct and indirect losses due to both to the natural and the cultural environment.

Ramifications of cyclone effects may be far reaching and depend largely on the ability of the country to buffer natural disasters by making available temporary shelter, food, medical aid and proper mitigative measures.

The managements plans include (a) risk assessment (b) timely ordered evacuation (c) government restrictions on insisting setback lines and building codes.

MANAGING SEA LEVEL RISE

It was been established that the mean sea level is increasing planning and investing for change can be introduced gradually within the constraints of finances, technology and man power. The potential economic and environmental damage from secular sea-level will lead to coastal erosion and redistribution of sediments, wet land submergence flood plain water logging and salt water intrusion into coastal aquifers. These changes will in turn cause disruption of residential, industrial and commercial activities, transport routes and ruin agriculutre for the posterity have been the talk of the day. Environmental modelling has become the need to achieve the above mentioned environmental targets. Man has to promote his understanding of the environmental systems and direct his own activities in consonance with natural laws to ensure his own future survival with a reasonable quality of life. Thus the broad scope of environmental modelling may be defined as follows:

(i) Discovery of the eco-dynamic principles

(ii) Estimate the potentials for sustainable exploitation of natural resources to promote the quality of life of man.

(iii) Conductor surveillance of his environmental interactions, with feedback analyses and modification of his actions.

Structure of a Model

A model is an abstract concept of the interactions among selected entities. For example, CO_2 emissions (sources), greenary which absorbs CO_2 (sinks) and carbonates in soil and water (sinks cum sources as per context) are projected as the entities that grovern the CO_2 level in air, which in turn has an influence on the global climate. A quantitative estimate of the resources and sinks to predict the global climatic changes is a conceptual model. A model has the following attributes.

(i) It is a simplified version of truth, with practical utility in predictions of phenomena. It is valid within a region of parametric variation and should not be stretched beyond.

(ii) It permits improvement with added knowledge and experience.

(iii) Certain universally accepted scientific principles underlie in a logical model. Every model needs calibration or validation by practical experience, check for internal logical consistency and conformity with established truths.

Quantitative models depend upon the following scientific principles :

(i) Mass and energy can neither be created nor destroyed.

(ii) Energy in all other forms tends to dissipate into useless heat.

(iii) Every transformation should result in entropy gain.

CLASSIFICATION OF MODELS

There are several approaches to environmental modelling, each with its own merits and demerits.

1. **Physical Models:** In these models the details of the prototype system are duplicated on a smaller scale. For example land erosion due to rain, easier to simulate and study. The changes in the behaviour of more during over crowding can be studied by crowding rats in closed space. Usually electrical and electronic systems are flexible enough to simulate natural phenomena of convenient modelling. Computers can simulate the global warming

due to Green house gases, taking several sources and sinks with complex dynamics. However linking the findings on the model to the prototype can have serious limitations. Exaggerated parallelism of the two phenomena can lead to bias misconceptions.

2. **Numerical Models:** The dynamics of an environmental system permits the framing of partial differential equations, which deal with the rate of change of one parameter with another etc. The differential equations need mathematical integration before they can be put to predictive use. But unfortunately several partial differential equations are not amenable use to integration. With the advent of computers, techniques for integrated giving approximate solutions could be developed. Such solutions are known as numerical models. At advanced levels of modelling, numerical solutions are the only hope since natural systems often display very complex dynamics.

ILLUSTRATIVE ENVIRONMENTAL MODELS

Growth of population with time: This model is called Exponential growth model, which is fairly valid when resources are plenty without any constraints.

$$dP/dT = KP$$

and

$$P_T = P_O e^{KT}$$

where, P_O = Initial population at any instant.

P_T = Increased population after time T and

K = Growth rate coefficient.

Exponential population growth model; Streeter Phelps' Oxygen sag curve for estimation of dissolved oxygen in a polluted steam and the Gaussian plume dispersion model for the estimation of concentration of plume at any point in space are some of the important environment models. Sustainable technological developments demand modelling the environmental parameters and their system dynamics. Modelling tools are diverse each with its own merits and limitations. Computer oriented numerical models carry the largest hope for environment modelling.

CHAPTER 7

Preserve Planet Earth

Earth was originated 5 million years back and is a planet in the solar system, it is only planet which supports the life. It is the third place from the Sun. It revalues around the Sun, while rotating around itself, with dolique axis, earth has about 78.5% of water and 25% of land. It has a special feature of possessing atmosphere which consists of gares like Nitrogen, Oxygen (inert gases are in minute quantities), which are responsible for supporting life. Life bring the beautiful Nature with seas, rivers, mountains, valleys and living kingdom which means plants and animals.

Man has to depend on nature for his survival. He also depends on both plants and animals for his existence. It is well known fact that Man is the top dominant trestious consumer in the food web. So, he gradually became dominant in nature ecosystem. By knowing importance of nature, people say that "Nature is God". Also a saying goes as Earth is Nature and Nature is beauty so only, "A thing of Beauty is a joy forever". Also in Atharvanaveda, it is said that, "O Earth, Whatever we dig:

"O Earth, Whatever we dig out from your, must that have to be filled up again and restored as fast as possible. O pure one, we don't intend to hit you at your heart of heart".

Which means that Earth is a sacred and a holy places created by God. Earth has its own natural system of maintaining. Ecological balance on Earth and creates a favourable environment for the living and non-living organisms. Though man injected some artificial systems of creaking and also destructing. Earth has its own super natural power of maintaining balance. The earth's atmosphere, which works as a giant heat uses engine uses the energy from the sun to drive the winds and upliftment water vapour from the ocean surfaces, forests and natural water courses to deposit in clouds which gives rain.

In short "Earth is the mother Goddess". She provides everything it gives energy and that is "shakti".

It includes:

1. Choice of compactable species in common habitat for better utilization at all tropic levels.
2. Reincorporation of organic elements resulting as by/waste product in the original form of the same system or other and
3. Mutual Co-ordination of systems.

Some continents of Earths parts are impregrable, its scenic beauty and picturesque scape and majesty to all of us lives. Alas, the beauty is waving life a meters. Flora and fauna of various species are at the verge of extinction. The ecological balance is threatened resulting in public mischief representing in regage of rivulets, erosion of valuable top layer of earth, extinction of ground water bed. The loss is incalcuble and rapid which is beyond man's capacity. The reasons can be assigned for this state of affairs.

Mans insatiable (unsatisfied) greed to over-rich and possess everything in nature even at the cost of depriving everything provided by matters bounty to the future of all living beings.

Lack of awareness in Environmental education and Protection. Hence environmental studies is a must to every individual.

With the technological and scientific advancement, in the quest to explore mineral wealth and complete in the market, the Government itself contributed for the indiscriminate exploration of mineral wealth by passing environmentalists, safety regulations and non-replacement of an upper layer of the earth compound the mischief to a great extant.

The proposed forestry programme for Earth will be aimed at accelerating greening of earth, conserving biological diversity, improving supply of goods and services in harmony with available natural resources and at providing better human environment to all.

In "Rigveda" it is said as ("O energy thou art the container of the entire cosmos since enternity. Thou art the bestower of sacrifice, knowledge and technology. O 'Respectable Ove' Thou art the source of light, seasons and all is beautiful. Thou art the creator and preserver of all the previous stones elements.

The earths energy that is shakti is innumerable, uncountable, infinite and is the supreme being. Shakti causes the great universal movements of the stars and planets and on our own earth, the great natural movements of wind and waters. Shakti is also at the orgin of five, heat and light and the power to enable men, animals and plants to grow and work.

Earth has its own good and bad it provides aim consumes i.e., it creates and destroys but it is maintained Ecologically balanced and environmentally favorable to all. As it is said good and bad exists side by side and so only we are able to appreciate the good. So earth almost provides everything to us but the question arises that as human beings what are we doing for the earth.

Are we creating, growing happiness and good will to earth by making good atmosphere or are we destroying Earth for our selfish needs, example by cutting trees, destroying forests, polluting the entire nature by with all artificial emenities. Making earth totally unclean and a polluted environment unfavorable for the living so by thinking upon earth the utmost we can do is to keep earth clean and neat.

Since "Cleanliners is Godliness".

There are 10 ways you can give Earth a chance.

1. Keep your car tuned.
2. When buying a new car make sure it has an antismog device.
3. Share a ride wherever possible.
4. Ride a bicycle or walk wherever possible.
5. Cutout or cut back the use of fertilizers, herbicides and pesticides.
6. Buy your milk in bottles.
7. Buy drinks in returnable bottles.
8. Do not depend on paper towels – use a sponge of towel.
9. Use linen napkins.
10. Use fewer electric appliances and fewer motor–run appliances only.

CHAPTER 8

Rio Earth Summit

Stockholms (Sweden) conference was held on 5 June, 1972 was the first Global conference to discuss Environmental issues in which 144 countries including India were participated. Our Late Prime Minister, Mrs. Indira Gandhi has attended the conference. This conference, though not a great success could atleast establish the global nature of environmental issues, which were earlier considered to be restricted to frontier economics of individual nations. G-7 Summit at Paris, 1989 identified pollution as a transboundary pollution.

Earth Summit at Rio De Janeiro, Brazil 1992, marked the beginning of a new era of cooperation between the rich and poor countries to save global environment. International conventions on Biodiversity and Climate were signed by majority of the nations attending the summit. Developing nations claim that 90% of the global environmental damage is due to the technological progress of the developed countries, who should therefore shoulder the major burden of the costs of global environmental protection. This conflict between the developed (Northen) countries and the developing (Southern) countries is still unresolved.

RIO EARTH SUMMIT

United Nations Conference on Environment and Development (UNCED), known as 'Earth Summit' or 'Rio Summit' was held for 12 days in June 1992 at Rio De Janeiro, Brazil. It was the largest non-religious meeting ever held and was attended by over 35000 environmentalists and other representatives and nearly 10000 journalists with the help of about 25000 soldiers and police.

It was represented by 170 nations and about 9000 ENGOs (Environmental Non-Government Organizations) participated in the Rio Summit in contrast to the Stockholm conference of 1972 in which majority were delegates from developed countries. Mr. P.V. Narasimha Rao, Prime Minister of India had the honour of opening the 3-day summit for Heads of State and Government at Rio. Maurice Strong, the Canadian environmentalist who chaired the 1972-Stockholm conference was once again chosen as the leader for this Summit also.

The twenty one (21) issues negotiated in the Earth summit were biodiversity, climate change, deforestation, environmental health, marine resources, ozone holes, poverty, toxic wastes, urbanization etc. Due to opposition by religions groups and racial minorities, the problem of population explosion was not discussed. The comprehensive 900 page document called 'AGENDA-21' gives the practical action plan giving the policies, laws, institutional managements and treaties to carry out the provisions of the different treaties and conventions and became crucial in ensuring environmental quality in the 21st century.

The agenda's 40 chapters are divided under four heads – social and economic dimensions, conservations and management of resources for development, strengthening the role of major groups, and means of implementation. The agenda is only a work plan without legal binding. A Sustainable Development Commission (SDC) was set up to integrate the activities of UN and agencies and to monitor the follow up on UNCED.

The deliberations of Rio summit could not resolve a conflict between developed and developing nations on two issues (i) Transfer of technology from developed to developing nations, needed for global protection was apprehended as detrimental to the intellectual property right of the developed nations, (ii) The flow of finances from the World Bank's Global Environment Facility was doubled to be directed to different end uses by the developing countries, other than environmental protective measures.

Chairman Strong estimated that the poorer nations need $125 billion per year from developed countries. Many environmentalists felt betrayed as the United States refused to sign the biodiversity treats, forest protection convention and also to donate new funds to poorer countries claiming that the conventions were too restrictive to American business and economy. Environmental Protection Agency Chief, William Reilly, was critical on George Bush, the then U.S. President for showing unwillingness to devote sufficient resources and time for the meetings. One year later, US president Clinton said that he would sign several treaties that former President George Bush rejected, agreed to offer $100 million for population stabilization and environmental protection and is now fully backing and encouraging the main theme of the summt, the 'sustainable development'.

Biodiversity Convention: The objectives of the Biodiversity convention of the Earth Summit are: Conservation of biological diversity, Sustainable use of biotic components. Fair and equitable sharing of the benefits arising out of the utilization of genetic resources, by appropriate access to genetic resources and by transfer for relevant technologies, taking into account all rights over those resources and to technologies, and by appropriate funding.

Climate Change Convention: CCC of the Earth Summit announced the following objectives: i) Stabilization of Green house gases at a level that would prevent dangerous anthropogenic interference with the climate system is to be achieved ii) Such as level should be achieved within a time frame sufficient to allow ecosystems to adopt naturally to climate change, to ensure that food production is not threatened and to enable economic development to proceed in a sustainable manner.

CHAPTER 9

Environmental Effects with Special Reference to Soils on Forests and Tribals

Out of 329 million hectares of the geographical area of India about 150 million hectares is at present under cultivation and more land is expected to come under the plough by 2000 A.D. Similarly the hydro-electric power forms one-third of the total generation of 55,000 M.W. The total hydro-potential of India is 85,500 M.W. at 60% load factor equivalent to 450 TWH of annual energy. The hydro power potential tapped in the South, West, East, North and North-East regions stands at 57%, 34%, 24%, 17% and 2% respectively. Most of the projects proposed to harness the surface waters in the rivers cater to both irrigation and power development. But the development of these projects must be on ecologically sound lines on sitting of River valley projects they become counter productive in the long run. Their environmental effects on forests and tribals is considered here.

Since most of the Civil and Electrical engineers design and construct dams and power houses as mere production units. Their harmful ecological impact is totally neglected while the engineers concentrate their attention on the optional methods of making available the intended benefits to the down-stream users, the adverse impact of the dams on the forests and tribals in the catchment areas upstream are ignored. The immediate victims are the tribals and the poor hill-folk whose lives depend on the resources provided by their immediate micro environment.

DAMS AND FORESTS

Although it is widely appreciated that the forest cover has to be restored 60% of land area in hilly regions and 20% in the plains with the ultimate object of raising the forest cover from its present level of 19.5% to 33% of the country's geographical area as per the National Forest Policy of 1952, little attention is paid to this crucial factor when a site is selected for a dam. As all the forests belong to the state, no compensation need to be paid to any individual. Moreover, in making cost, the forest loss is grossly under valued so that cannot confirm to the guidelines to the planning commission approval for a benefit to east ratio of 1:1. During the construction of the dam, the forests suffer in many ways. In the case of dams in Himalayan valleys, the contractors men not only cut the forest that will be submerged but also the useful trees even in unaffected areas. In the case of Tehri Dam, the people from the old Tehri town are shifted from the areas to be submerged and rehabilitated on forest lands that are acquired and cleared in the neighbourhood of Rishikesh and Haridwar. The authorities acquired good forest land for formation of roads and construction of houses and offices in the New Tehri township area. With the arrival of people, there will be an increasing demand for wood for fuel and house-hold purposes and most of the remaining forests will be slowly denuded and hence construction of a dam is bound to have a multiplier effect in reducing the area under forests. In fact Iddukki dam over Periyar river in Kerala hastened degradation of vegetation, reduction of forest cover and weeds of lantana found their way into the valuable forests. About 0.5 million hectares of forest was lost due to major dams during 1951-1976.

The ecological significance of forests was brought into focus by the environmentalists who organized a national campaign against the Silent valley project in Kerala. Although some ecologically illiterate politicians looked at the problem from the relative importance of monkey versus man, the environmentalists rightly focused the threat of extinction posed by the dam to the priceless biological wealth that evolved over hundreds of millions of years. A preliminary study indicated 900 species of flowering plants and ferns and a few plants are new to science itself. The vegetation consists of many medicinal and aromatic plants as also the wild relatives of spice and pulse-yielding plants, that have wide genetic diversity with respect to protein content and disease-resistance lost. Among the uniques birds, insects, fisher reptiles and mammals in the Silent valley, the lion-tailed Macaque is one of the most endangered mammals in the world. While Mr. Moraji Desai, the Janata Prime Minister of India cleared the Silent valley project. Mrs. Indira Gandhi, a champion of the environment helped in shelving the project in 1983. Vast stretches of forests in Himalayas, Western Ghats and Eastern Ghats have been lost due to irrigation projects. Major hydel projects like Koyana, Sharavathe, Bhavanic and Iddukki projects ruined good forest areas in Western Ghats. Some of the reservoirs at Panshet, Kabini, Kali and Bhavani Sagar degraded the biological diversity in different forest ecological systems. The proposed Tehri dam on Bhagirathi in Tehri Garhwal region of the Himalayas. A number of hydro electric projects are proposed in the Himalayan catchment areas of the Ganges and Yamuna and their tributaries and this programmer is bound to convert the Himalayan region into biological deserts. Large-scale deforestration promotes the colonization of the denuded areas by weeds such as *Eupatorium glandulosum*. Such weeds become unpalatable to the herbivorous animals whose ecological niche will be uninhabitable. Denudation renders the forests highly susceptible to forest fires and to mineral losses through leaching. Impounding the river waters in the upper reaches diminisher the normal river flows to almost a trickle and consequently the dilution and dispersal of pollution will be adversely affected and hence the quality of water deteriorates and the local eco systems get damaged. The storage of water in the reservoirs behind the dams is often reported to cause Reservoir induced Seismicity and it may increase the frequency of tremors as well as earth quakes. As a consequence of the unstable mountain slopes of Himalayas may collapse resulting in large scale destruction of forests. The construction of major dams like Nagarjuna Sagar and Srisailam on river Krishna in Andhra Pradesh has converted many forests lands into barren areas. Similarly, the construction of major reservoir, Mettur dam in Tamil Nadu and Tungabhadra and Machkund in Andhra Pradesh have drastically reduced the forest cover and consequently altered the nature of some of the unique ecological systems in the Eastern Ghats. The planning of major dams in the highly seismic areas of Assam and its environs will spell ruin to the forests.

TRIBALS

The construction of several large scale dams results in storage of huge amounts of water which submerge man's villages and forest lands in the catchment. The gigantic Srisailam dam on river Krishna in Andhra Pradesh is reported to have submerged over one lakh acres of land including forests, displeasing about one lakh population with 80% of them being scheduled and backward caste people. The Government officials are least bothered to preserve and create the community life of the displaced people. Such displaced people are treated as a collection of unrelated individuals in a mechanical outlook that characterises the administrative set up of pre-independence of India. The refugees are some times paid compensation in cash which is usually spent on liquor and meet other expenses of the family with the result that they loose their traditional methods of taking out their livelihood by tilling their ancestral lands. The sudden change of the refugees to a new social order or modernization, the destruction of their natural environment on which their mental and physical health as also their life style was dependent, and resettlement problems make a tradition-bound tribal family totally perplexed powerless and puts them on the verge of total social cultural and economic collapses. When works on Panshet dam started in 1958, the people to be displaced offered Satyagraha and went to jail, and demanded land for land in the command area benefitted by the dam. As a result of prolonged agitations by the people displaced by dams, the Maharashtra Government passed the Maharashtra Resettlement project displaced persons Act 1976, "that" grants the refugees land in the command area. While the farmers get at least half of the land in the command area. In the case of Tehri dam, the oustees are given 2 acres of developed agricultural land to each displaced family or 0.5 acre land within 5 km of Dehradun city in lieu of acquired land 0.2 acres of land to landless labourers is also provided free of cost. As per the Land Acquisition Act, cash compensation at a minimum of Rs. 40,000 plus exgratia payment of Rs. 12,000 per acre for irrigated and Rs. 4000 to Rs. 6,000 for unirrigated land is paid Scheduled castes and tribes landless labourers get 2 acres of land.

It is a pity that the immense social, economic and cultural costs to be borne by the oustees and others affected by the development projects are dismissed by the bureaucrats as in inevitable cost of development. If Tehri dam displaces over 90,000 people and loose 5200 hectares of land, the Narmada project renders rootless over 2 lakh people besides 13,744 hectares of forests and 11,318 hectares of agriculture land. Since Gujarat state has to bear the costs of rehabilitation of the oustees from Maharashtra and Madhya Pradesh, the Narmada tribunal award insisted on implementation of Rehabilitation measures like, "land for land", and basic amenities like drinking water, schools, hospitals etc. in the case of Sardar Sarovar project. The Narmada tribunal specified that a minimum of 5 acres should be given to each affected family. Shifting tribals from uplands of Inchampalli and Bhopalapatnam and rehabilitation them in an alien Marathi speaking area amounts to "ethnocide" in the words of the social activist, Baba Amte. The Dihang hydel project estimated to produce 20,000 MW of power with a 300 meter dam across The Siang river drown about 500 sq. km. of fertile land including 91 villages and 3 towns. The local tribal leaders are against the project as it would affect the lifestyle, culture and environment of the ecological sensitive area. In rehabilitating to the Tribals from the catchment area of major development projects, the following 5 principles of tribal development as accepted during Jawaharlal Nehru's time, known as "Tribal panch sheel" must be followed:

1. Tribals should develop along their own traditional genius and life-styles and nobody should impose anything on them from outside.
2. Tribal culture and traditional arts should be encouraged.
3. Tribal rights in land and forests must be preserved.
4. Tribals should be trained to run the administration and development programmes.
5. Outside agencies should work through and not in rivalry with the tribal social and cultural institutions.

As far as possible, all development projects should be planned so that there will be no clash between the environment and development.

(*Sources:* State of India's Environment, 1982-1984)

Chapter 10

Environmental Education, Monitoring and Soil Managements

As concern for environmental protection and conservation of environment have increased world-wide, the detrimental impacts of pollution and human dominance on natural ecosystems have becomes increasingly important for resource management. There is no doubt about the progress and prosperity brought about by industrialization, however, the materialistic progress posing has ruthlessly demolished the very natural surroundings and is posing has ruthlessly demolished the very natural surroundings and is posing threat to environment. Deforestation has led to destruction of wildlife, soil erosion, loss of soil fertility, floods and formation of deserts. Urbanization and improper planning expansion of cities have created the problem of waste disposal, sanitation and provision of pure water and clean air, use of synthetic chemicals, as pesticides in agriculture are also intensifying the problem of water pollution. Pollution of water, air and soil is formidable problem at globally.

The environmental problems of developing countries are not just the side effects of excessive industrialization and also due to man's incomplete knowledge of the possible impact of scientific development. Progress has become synonymous with an assault an nature. Industrialized countries witnessed much more acute environmental pollution and began realizing its adverse impact of the lives of the people, which were sought to be enriched through industrial production. The United Nations conference on Human Environment held at Stockholm in June, 1972 proved very timely with it began an era of increasing consciousness world timely with it began an era of increasing consciousness worldwide regarding the urgent need for protection of the Environment.

It was fortunate that our country took up this matter in right earnest. To make people aware of Environmental protection constitution was amended. Article 48(A) & 51(A), (G) were added to the constitution there by imposing duties on Government as well as citizens for protection and improvement of the environment.

STRATEGIES OF ENVIRONMENTAL MANAGEMENT

Environmental management is a process which, by taking into account the overall ecological, cultural, economical, social, technological and other factors, attempts to ensure that the human environment is developed in an integrated and systematic manner. It aims at the best use of existing and potential resource base, in such a way as to maintain for now and in the long term, the best possible sustained yield from the biosphere. Its ultimate objective is to provide greater personal and social opportunities and to improve overall human well being for present as well as future generations. Some of the objectives of environmental management are:

- Environmental planning
- Environmental status and Evaluation
- Environmental Impact Assessment
- Environmental Legislation and Administration

To be successful, environmental management requires first the availability of particular parameter data to assess its present status. Data can be obtained by a monitoring system designed to measure at different points of space and time, various selected parameters and these can be consolidated into appropriate indices which characterize the quality of environment. Secondly, to identify the consequence due to alteration in the environment quality, on existing both biotic and abiotic components of ecosystem including human response do it. The third action involves the projection of the proposed action and the presentation of the results in a manner that the analysis of probable environment consequences can be incorporate or used in the decision making process.

Government of India in 1972 constituted the National committee on Environmental planning and coordination (NCEPC), some of the objectives of the committee were to advice on environmental problems and made recommendation for their improvement. To carry out these objectives, the NCEPC is to review, formulate and promote policies and programmes covering development projects, Physical Planning Legislation, administrative procedures, education and public information and research. Department of Environment was created with effect from November 1980 under the charge of Prime Minister, its functions are promoting research in environmental problems, pollution monitoring, conservation of natural resources and marine ecosystems. Various plans have been formulated for environment management in India. Now it is MOEF (Ministry of Environment and Forest, Govt. of India), known as "Paryavaran".

PLANS FOR CONSERVATION OF FORESTS

It is clear that for the last long years there has been vast decrease in deterioration and denudation of forests. Forests are one of the valuable natural resources and they are very basis of our survival as to maintain in the proper balance of organisms, chemicals in the environment by producing or absorbing them and preventing soil erosion, floods, environment, droughts, etc.

The Government of India revised the old forest policy of 1952 in December 1988. The major objectives of the new National Forest Policy are:

1. To maintain ecological stability by adopting appropriate measures of environmental protection, conservation and restoration,
2. To implement social forestry project on large scale,
3. Protections of existing floods,
4. Protection of Wildlife and Biodiversity,
5. To meet basic requirement of rural people and tribal for fuel, fodder and shelter purposes.

Protection of forests cannot bet possible unless a person volunteers himself or herself to cooperate. Therefore, it is necessary to create awareness about protection and development of forest among the people. (*e.g.* Chipko movement in Himalayas) and Apiko movement in Karnataka state).

ENVIRONMENTAL PROTECTION, WATER AND AIR POLLUTION

Water and Air are important for life. Control over sewage and industrial discharge into rivers and disposal of waste is so fragile that all most no pollution control is possible. Air pollution has become a great menace to life and it causes severe health hazards. For the control of water and air pollution there are three main acts.

- The Water (Prevention & control of pollution) Act, 1974
- The Air (Prevention & control of pollution) Act, 1981
- The Environmental (protection) Act, 1986

The Government agencies must play a constructive role to provide guidance and advice to the industry in understanding the requirement of the laws and ensuring their faithful compliance.

CONCLUSION

Present methodology of environmental management requires the conscious generation of alternatives at different levels of planning (i.e., at the policy, programme as well as the project stage). Afforestation programme implemented should be at local. There is a need for reorganization of Pollution boards, to make people aware of environment protection, conservation and Environment education should be made compulsory for childhood level.

ENVIRONMENTAL EDUCATION

Since prehistoric time, man developed certain tools and techniques to fulfill his basic needs. But time to time changes in the human society, civilization and technological development. Man is the creator and consumer of so many things. Now the increased unethical activities of man are responsible for the present state of environmental crisis.

The pressure on resources increased with the increasing rate of population. The rapid industrialization, urbanization, deforestation, desertification, threatened wildlife nature, over grazing shifting cultivation, dwelling process, soil erosion, climatic changes, increased rate of pollution in air, water and land. The loss of Biodiversity and so on are responsible for the present environmental disasters like "Bhopal Gas Tragedy". "Chernobyl incident" and other accidents threatening the mans life.

The following are a few examples roughly to estimate the environmental crisis:

1. At present the world forest cover is decreasing by 18-20 million hectares per year. India is losing 1.5 million hectares of forest cover in every year. The recent Satellite survey is showing the forest cover is not more that 10% in the total geographical area in our country.
2. It is estimated that the flood damage cost is near about Rs. 1,000 crore per year in the country.
3. Soil erosion is taking place at an accelerated phase, more than 6000 million tons of fertile top soil is eroding every year. The cost in the economic point of view is more than Rs. 12,000 crore.
4. Desertification is also taking place at an accelerated phase during the last 50 years, 65 thousand million hectares of productive land in southern parts of Sahara region has converted into desert and nearly 14% of the word's population is living in the threatened dry lands.
5. The use of "non-renewable" resources, such as coal, petroleum, minerals etc. has increases in recent times.
6. The loss of "Biodiversity" will effect the balance ecosystem.
7. Pollution causing severe damage to the environment.
8. Climatic changes are expected in the early part of 21st century.
9. Environmental refugees are more and more.
10. Effect of sustainable development and so on.

The above mentioned issues are only examples. Besides these we have so many environmental problems. Now the question with us is how to over come these problems and what can be done to resolve these issues? The only good approach is promoting Environmental education at all levels including primary school/ college levels.

Environmental education is to promote an understanding about the sensitive nature and man's interaction with physical, chemical, biological, economic, political, social, cultural aspects of environment and to awaken human societies to safeguard the only life supporting system in the universe.

Environmental education is a process to bringing awareness, knowledge, skill, development, attitude, motivations and commitments to work individually and collectively to solve the present environmental problems and to prevention of new one's.

The Biosphere is the central theme of environmental education to build up environmentally literate citizens with responsibility. The environmental education and curriculum should be based on the present environmental trends and should reflect the man inter-relationship with nature. It must be a remedy to measure the present situation.

Environmental education at global level was discussed in the Stockholm conference in June 1972 and then Belgrade Charter in 1975. The first International conference on Environmental education held in Tbilisi, 1977. The first environmental education conference in India held at New Delhi in 1981. The International workshop on development and Biosphere stability, New Delhi in 1985 and adopted a declaration to protect the nature and stressed the need of "Environmental Awareness.

ENVIRONMENTAL AWARENESS THROUGH

1. Both formal & non-formal education.
2. National parks, Sanctuaries, Botanical, Monuments etc.
3. Through future article, Interviews, News time.
4. Newspapers.
5. Radio & Television.
6. Slogans, Cartoons, Drawings etc.
7. Awareness campaigns, field visits, demonstrations, meetings, seminars, exhibitions.
8. Books, Journals, Reports, etc.
9. Charts, Posters, Films, Slides etc.
10. Through Folk media.
11. Organizing Eco-camps.
12. Through Science clubs, essay writing, debate, drama and other extra curricular activities.

The following organizations working at International level to promote Environmental education and to conserve nature. The WWF, ISEE, NAEE, IUCN, WCMC, FAO, SCOPE, GEF, FEC, WMO, WHO, SACEP, EnRA and other organizations were involved in the field of environment.

In India more than 908 Non-government organizations including EnRA are working in the field of environment. The union Government has launched NEAC campaign to build up Environmental awareness. Finally the peoples participation is very important for the success of any programme. Hence the approach should attract the people to involve in the environment education programme for quick results.

ENVIRONMENTAL MONITORING AND ENVIRONMENTAL ISSUES

Environmental monitoring is the systematic collection of physical, chemical, biological and related data pertaining to environmental quality, pollution sources and other factors that influence or are influenced by environmental quality. Environmental quality data are vital components for one or more of the following uses:

- Establishments of or revisions to standards.
- Demonstration that adequate progress is being made for that attainment of the standards, in the scheduled time frame.
- Providing assurance that compliance with standards has been attained.
- Information regarding maintenance of standards.
- Determine if control of high pollution episodes or spills is adequate and provide guidance on choice of actions.
- Definition of environmental pollution problems for periodic determination of priorities for resource allocations and the development of control programs.

OBSERVATIONS

The main goal of the monitoring program is designed to sample the environment for pollutants for which national environmental standards has been promulgated. An example of such standards is the National ambient air quality standards for particulate matter, Sulfur dioxide, Nitrogen dioxide, Carbon monoxide and Photochemical oxidants.

Environmental monitoring related activities are divided into the following categories: Trend monitoring, Ambient-source linked monitoring and exposure monitoring. Monitoring of trends involves the measurements of pollutants or the effects of pollutants in air, water, soil and biological matter over extended periods of time. Ambient source linked monitoring involves relating the ambient environmental quality to sources through modelling, considering other pertinent supporting data such as meteorology, hydrology, demography, topography and so on. Monitoring exposure involve the measurement of pollutant concentrations in air, water and food at the location where exposure may occur.

Table 10.1: Atmospheric Pollutants Measured by EPA

Element	*Radicals*	*Gases*	*Others*
Antimony	Ammonium	Carbon monoxide	Aeroallergens
Arsenic	Fluoride	Methane	Asbestos
Barium	Nitrate	Nitric oxide	Benzene-soluble
Beryllium	Sulphate	Nitrogen dioxide	organic compounds
Bismuth		Ozone	Benzene (a) Pyrene
Boron		Pesticides	Pesticides
Cadmium		Reactive	
		hydrocarbons	Radionuclides
Chromium		Sulphur dioxide	Respirable particulars
Cobalt		Total hydrocarbons	Total suspended
			particulates
Copper			
Iron			
Lead			
Manganese			
Mercury			
Molybdenum			
Nickel			
Selenium			
Tin			
Titanium			
Vanadium			
Zinc			

The present day monitoring of the nations environmental quality is a cooperating effort involving local, state and federal agencies as well as industry. Most monitoring effort are in the area of air and water. There are also monitoring programs such as the National Pesticide Monitoring Program. The atmospheric monitoring program conducted by local and state agencies is directed toward those pollutions for which National Ambient Air quality standards have been promulgated, namely particulate matter, sulphur dioxide, nitrogen dioxide and photochemical oxidants. Table 10.1 provides a list of atmospheric pollutants currently being measured by EPA (Environmental Protection Agency).

Water quality monitoring is also carried our by many different agencies of groups. Most local municipal water treatment facilities monitor raw water quality daily. Many municipal wastewater treatment programs and country agencies also routinely monitor receiving waters upstream and downstream from treatment plant discharges. Many universities regularly collect water quality data. Table 10.2 shows a list of parameters presently used in the evaluation of water quality.

Table 10.2: Parameters used in the evaluation of the Water Quality

Dissolved oxygen, pH, Coliform, Temperature, Floating solids, settlable solids, Turbidity or Colour, Taste-Odour, Toxic substance, Radioactive substances, Total dissolved solids, Methylene blue active substances, Zinc, Salinity, Chlorophyll, Arsenic, Barium, Cadmium, Chromium (hexavalent & trivalent), Lead Selenium, Silver, Suspended solids, Chloride, Copper, Nitrate, Phenols, Phosphate, Sulfate, Cyanide, Electrical conductance, Ammonia, Acidity, Alkalinity, Carbon chloroform extract, Fluoride, Hydrogen sulphide, Pesticides, Sodium, Iron, Plankton, Foaming substances, Boron, Manganese, Hardness, Biological Oxygen Demand.

MONITORING SENSORS AND METHODS

Assessing the quality of the environment depends to a considerable degree upon sensor availability and applicability. Until recently, most of the available pollutant sensors capable of providing quantitative or qualitative information were in situ or contact type instrumentation. Such sensors are restricted to assessing some chemical, biological or physical parameters of the environment at a specific point or when mounted on a mobile platform, at sequential points as a function of time. Presently the use of remote sensors came into existence, which are capable of very rapid measurement of an array of points in space and time. The use of remote sensing is not a new concept. Many federal agencies, including the EPA, the National Aeronautics and Space Administration and the Department of Transportation have sponsored extensive development of such sensors. Initial work in air monitoring concentrated on measure of temperature, water vapour and ozone primarily from satellites using the upwelling infrared radiation from the earth. Photographic techniques have been refined in the water and land media. They provide information on location of pollution outfalls, extent of oil spills, vegetal stress, impact of urbanization on water bodies and the presence of sedimentation or other water plumes.

An example of such remote system developed by the environmental monitoring and support laboratory – Las Vegas, which has recently become available for field use is the air craft mounted light intensity detection and ranging system (LIDAR). This instrument uses a pulsed laser and telescope arrangement to detect and range atmospheric particulate matter. The laser mounted on an aircraft flying 10,000 feet is periodically fired toward the ground. The telescope collects the returning reflective signal from the ground and from particular matter in the air and the information's stored on tape for subsequent analysis. There are also present individual exposure sensors. These have only been perfected in a practical sense in the field of radiation. These devices, placed on an individual measure his integrated exposure to specific types of irradiation regardless of his time-spatial relationship to source. Such devices are developed to measure the integrated exposure of an individual to a variety of air-borne pollutants.

CONCLUSIONS

Historically, environmental monitoring has been conducted more in a reactive fashion than in a planned systematic way. Furthermore, regardless of the purposes for which the monitoring began, there is usually considerable resistance to change any existing network once a substantial body of has been collected. It is now time to re-evaluate all existing environmental monitoring from a more systematic and analytic point of view.

Appendices

APPENDIX 1

SCIENTISTS' PRESCRIPTION TO MAKE SOIL FERTILE

Diagnostic teams recommend application of Zinc Sulphate, Andhra Pradesh at Visakhapatnam:

A study of quality of agriculture land soils by scientists has revealed sulphate deficiency where paddy, oil seed, and maize crops are being raised. The nutrient uptake is hampered, affecting the productivity of the crop adversely in the district.

The deficiencies have been observed generally in paddy nurseries and lesser in transplanted paddy. The symptom is observed in pulses and oil seed cultivation lands. The diagnostic teams of research scientists and extension officers, who visited the crop fields in the district, have recommended application of Zinc Sulphate in soils as well as foliar spraying wherever deficiency is noticed.

The scientists have recommended Zinc application at 50 kg per hectare in two to three crop seasons without fail to enrich the soil and turn it fertile. The application should also be extended to waste lands, which are being converted into crop lands. Zinc should be applied in ill drained and problematic soils, including alkaline soils, where organic carbon is more than 71.5 per cent in the organic matter. Application of Zinc is also recommended in the hilly areas during rabi season, as Zinc deficiency is glaring under cool temperatures.

Past Experience

Agriculture officers monitoring the Zinc deficiency fields in the district are of the view that experience in the past couple of years has proved that application of Zinc Sulphate resulted in increased production and productivity of paddy, apart from its quality. The exercise is highly beneficial to the farming community. As per the Government guidelines, nodal agencies supplying zinc sulphate to farmers should procure 21 per cent of ISI Zinc Sulphate from type manufacturing companies. The dosage recommended is 50 kg per hectare for paddy crop and 25 kg per hectare for groundnut crop. Supply of Zinc to beneficiaries should not exceed two hectares. The beneficiaries covered during the past three years should be ignored and those not covered under the programme sofar should be given priority.

Lab Analysis

The Mandal Agriculture Officer has to collect samples of the chemical supplied and send the same to the laboratory for analysis of its quality. The test results should be given within 48 hours. Stocks should not be distributed until the Agriculture Officer attests its quality.

The Joint Director of Agriculture would constitute teams for verification of distribution of the chemicals. In the 2011-12 financial year, 1426 hectares of crop lands were covered in the district. Currently, an action plan is being drawn to bring more hectares of crop lands under the Zinc Sulphate programme, surpassing the last year's target.

APPENDIX 2

LAND IN INDIA

Of all our resources, land is the most tangible one. Land, by definition, is the solid part of the earth's surface. It is finite resource, so great care should be taken no preserve it. Land has been put to many new uses, apart from traditional ones.

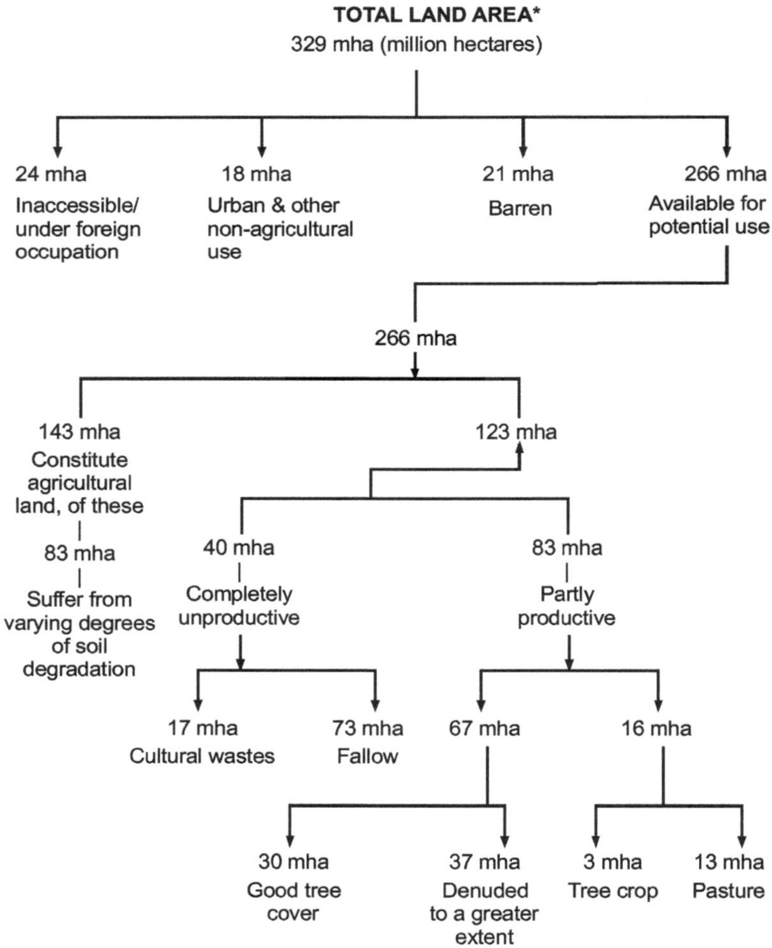

These are the general uses of land:

- Housing-soil for construction use
- Laying of roads and railway lines
- Industries
- Mining of mineral resources
- A container for holding water
- Traditional uses like agriculture for food and fodder, cattle grazing, forest wealth for timber, fuel etc.

Sharing of 1 km^2 of Land by No. of Individuals = 10 individuals

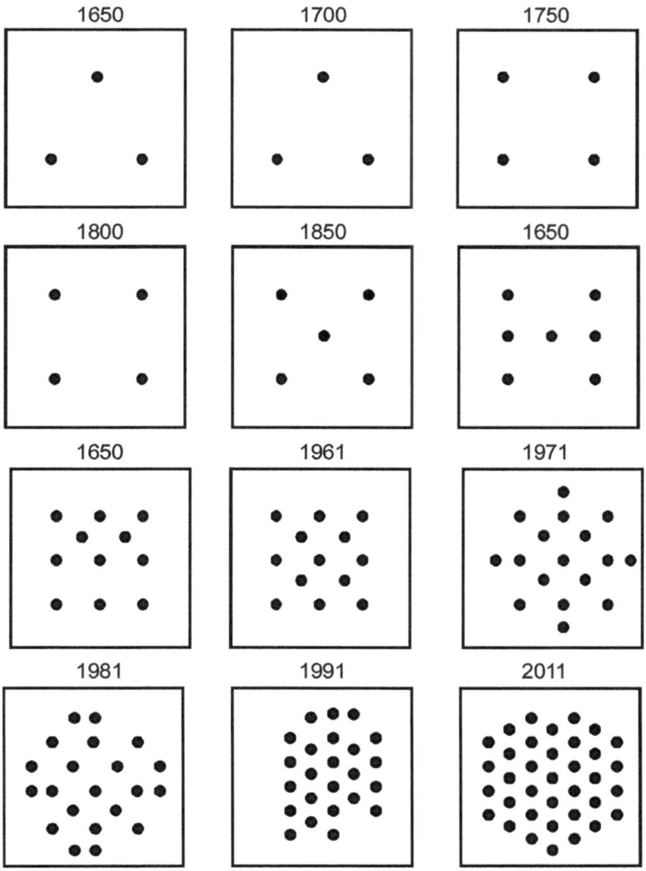

APPENDIX 3

ENVIRONMENTAL SUPERLATIVES

Mountain Peaks		
1. Mount Everest	Nepal-Tibet	8708 m
2. Kanchanaganga	Nepal-India	8484 m
3. Mt. Godwin Austin (K2)	India	8475 m
4. Nanda Parbat	India	7998 m
5. Annapurna	Nepal	7951 m
6. Nanda Devi	India	7694 m
Great Deserts		
1. Sahara	N. Africa	8960,000 sq. km.
2. Libyan	N. Africa	1664,000 sq. km.
3. Australian	Australia	1536,000 sq. km.
4. Gobi	Mongolia	1024,000 sq. km.
5. Thar	N.W. India	256,000 sq. km.
Islands		
1. Greenland	Arctic Ocean	2150,000 sq. km.
2. New Guinea	West Pacific	812,000 sq. km.
3. Great Britain	North Atlantic	216,000 sq. km.
4. New Zealand	S.W. Pacific	148,000 sq. km.
5. Sri Lanka	Indian Ocean	65,000 sq. km.
Rivers		
1. Amazon	S. America	6712 km.
2. Nile	Africa	6632 km.
3. Mississippi Missouri	USA	5969 km.
4. Yangtze	China	5440 km.
5. Ob-Irtysh	USSR	5120 km.
6. Indus	Asia	2720 km.
7. Brahmaputra	Asia	2688 km.
8. Ganges	India	2640 km.

9. Colorado	USA-Maxico	2320 km.
10. Peace	Canada	1912 km.
11. Darling	Australia	1856 km.
12. Snake	USA	1661 km.

Lakes

1. Caspean Sea (Salt)	USSR & Iran	367,000 sq. km.
2. Lake Superior (Fresh)	US & Canada	81,500 sq. km.
3. Victoria Nayana (Fresh)	Canada, Tanzania, Kenya	68,500 sq. km.
4. Aral Sea (Salt)	USSR	65,000 sq. km.
5. Lake Hurom (Fresh)	Canada & US	59,000 sq. km.

APPENDIX 4

ADDRESSES OF INTERNATIONAL ORGANIZATIONS/NGOS/FUNDING AGENCIES

Dealing with issues of Ecology and Environment

1. NIRMA INSTITUTE OF TOWN PLANNING
   Near Chinmay Towers, Subhash Chowk, Ahmedabad, India.
2. CENTER FOR ENVIRONMENT AND TOWN PLANNING (CEPT)
   Navarangapura, Ahmedabad, Gujarat, India
3. DELHI SCHOOL OF PLANNING AND ARCHITECTURE
   Indiraprastha Estate, New Delhi, India.
4. EARTH DAY RESOURCES
   116, New Montgomery, Street, Suite 530, San Francisco, CA 94 105 USA
   (Maintains a network of environment-based organizations and offers materials like lesson plans, recycling and ride-share programmes).
5. EARTH WATCH
   P.O. Box 403, Watertown, MA 02272, U.S.A.
   (Links volunteers with scientists and scholars for academic research expeditions)
6. ECO-NET
   18, Deboom Street, Dept. GM, San Fransisco, CA 94107, USA
   (Deals with environmental isssues and Job opportunities: a computer network)
7. GLOBAL ACTION PLAN
   84, Ferry Hill Road, Woodstock, NX 12498
   (Provides support for formation of "Eco Teams" and releases a workbook)

8. PUBLIC INTEREST RESERACH GROUPS (PIRG)
   29, Temple Place, BOSTON, MA 02111, USA
   (Provides environment related canvassing and fund rising jobs for young graduates)
9. STUDENT ENVIRONMENTAL ACTION COALITION (SEAC)
   P.O. Box 1168, Chapel Hill, NC 27514, USA
   (Sponsors students interested in Environmental and Ethics)
10. CARE
    660, First Avenue, New York, NY 10016, USA
    (Offers mechanical and material assistance to developing countries for potable water, health, nutrition, environment, emergency, agriculture etc.)
11. CENTRE FOR ENVIRONEMNT EDUCATION (CEC)
    1725, DeSales Street, NW, Suite 500, Washington, DC 20036, USA
    (Sponsors programmes for protection of endangered species)
12. CLEAN WATER ACTION PROJECT (CWAP)
    1320, 18th Street, NW, Washington, DC 20009, USA
    (Works for clean water and protection of natural resources)
13. RESOURCES FOR THE FUTURE OR ENVIRONMENTAL LAW INSTITUTE
    1616 P, Street NW, Suite 200, Washington, DC 20036, USA
    (Sponsors Research and Education on Environmental Law and Policy)
14. INSTITUTE FOR SOCIAL ECOLOGY
    P.O. Box 89, Plainsfield, VT 85667, USA
    (Offers workshops of Environmental based subjects)
15. INTERNATIONAL UNION FOR CONSERVATION OF NATURE AND NATURAL RESROUCES (IUCN)
    Avenue du Nont-Blac, CH-1196 Gland, Switzerland
    (Promotes scientifically based action for conservation of wild animals and plants)

16. NATIONAL WILD LIFE FEDERATION
    1400, 16th Street, NW, Washington, DC 20036, USA
    (Its over 5 million members work for wildlife conservation protection)
17. PLANET/DRUM FOUNDATION
    P.O. Box 31251, Sanfrancisco, CA 94131, USA
    (Promotes new attitudes towards nature & wise use of resources)
18. UNITED NATIONS ENVIRONMENT PROGRAMME (UNEP)
    P.O. Box 30552, Rm DC2-0803, New York, NY 10017, USA
    (Coordinating global environmental issues with UN Agencies, Govt. & NGOs)
19. WALDEBRIDGE ECOLOGICAL CENTRE
    Worthyvale Manor Farm, Camelford, Cornwall, P 132 9TT, England
    (Research and education on environment, publishers 'the Ecologist', an international journal)
20. WORLD WILDLIFE FUND (WWF)
    1250, 24th Street NW, Washington, DC 20037 USA
    (Protection of world's Wildlife and Habitat)
21. ENVIRONMENTAL PROTECTION AGENCY (EPA)
    401 M Street SW, Washington, DC 20460, USA
    (A Govt. Agency that enforces clean air and water laws)
22. COMMUNITY AID ABROAD (CAA)
    Fitzory – 156, George Street, Fitzory VIC 3065, Australia
    (Water Development and Sanitation)
23. CONSERVATION, FOOD & HEALTH FOUNDATION
    P.O. Box 185, Wellesley Hills, 02181-6824, Massachusetts, USA
    (Environment, Wildlife and Global relief)
24. CURTIS & EDITH MUNSON FOUNDATION
    P.O. Box 3488, Cleanwater Beach 34630, Florida, USA
    (Environment, Wildlife and Conservation)

25. EARTH SCAN
    3, Endsleigh street, London, WCIH ODD (UK)
    (Environmental Protection and Preservation)
26. EARTH WATCH
    C/o UNEP, P.O. Box 30552, NAIROBI, KENYA
    (Environmental Protection and Preservation)
27. THE FISHER FUND
    85, Broad Street, 2nd Floor, New York 10004, USA
    (Animal welfare, Environment and Education)
28. BELGO INDIA VILLAGE RECONSTRUCTION ORGANIZATION
    Pastoriji Street 16, 9360, BEGGENHOUT, BELZIUM
29. BELLERIVE FOUNDATION
    Case Possale 6, 2011 Geneva 3, Switzerland
    (Self conducted environment-based programmes)
30. THE BYDALE FOUNDATION
    750, 7th Avenue, 24th Floor, New York 10019, USA
    (Social service, education, environment)
31. COALITION FOR WOMEN IN INTERNATIONAL DEVELOPMENT (CWID)
    C/o Overseas Education Fund, 2101 L Street, N.W.,
    Suite 916, Washington, DC 20037, USA
    (Women and Environment, Women Development)
32. INTERNATIONAL YOUTH FEDERATION FOR ENVIRONMENTAL STUDIES AND CONSERVATION
    Klostermile, Klostermllevej, 48A, DK-8660, 5 KANDERBORG, DENMARK
    (Environmental Protection and Preservation)
33. THE NEEDMORE FUND
    1730, 15th Street, Boulder 80302, Colorado, USA
    (Environment, Rural and Community Development)
34. WORLD WILDLIFE (WWF)
    Panda House, 11-13, Oxford Road, Godalming, UK
    (Environment, Wildlife, Conservation)

35. CENTRE FOR ENVIRONMENT MANAGEMENT & PLANNING
    Auris Business Centre, 23 Street, Machar Drive,
    Old Albedeen, AB1 IRY, Scotland, UK
    (Training Centre in Environmental Management)
36. UNITED NATIONS DEVELOPMENT PROGRAM (UNDP)
    1, UN Plaza, New York, USA
    (Environmental Management)
37. INSTITUTE TERRESTRIAL ECOLOGY
    Banchery Research Station, Hill of Brathers,
    Glassel, Banchory, Kincardin Shire AB 314 By Scotland (UK)
    (Environmental Education)
38. FOUNDATION FOR ENVIRONMENTAL CONSERVATION (FEC)
    7, Chemin Traverney, 7th & 8th Floors, 1218 Grand-Saconnex, Geneva, Switzerland
    (Environmental Protection and Preservation)
39. FRIENDS OF THE EARTH
    26-28, Underwood Street, London N1 JQ, UK
    (Environment and Community Development)
40. UNESCO
    Regional Office of Science & Technology for South and Central Asia, 8th, Poorvi Marg, Vasant Vihar, New Delhi-110057, India
41. SOCIETY FOR PROMOTION OF WASTELAND DEVELOPMENT
    Shriram Bharatiya Kala Kendra,
    1-Copernicus, Marg, New Delhi-110001, India
42. CAPART
    India Habitat Centre, 2nd Floor, Zone No. 5,
    Lodhi Road, New Delhi-110003, India
    (Rural Development)
43. NATIONAL INSTITUTE OF EDUCATIONAL PLANNING & ADMINISTRATION (NIEPA)
    NCERT Campus, Sri Aurobindo Marg, New Delhi-110016, India
    (Environment, Education)

44. NORWEGIAN AGENCY FOR DEVELOPMENT CORPORATION (NORAD)
    Royal Norwegian Embassy, Shantipath, Chankyapuri, New Delhi – 110021, India
    (Environmental Awareness, Women Development)
45. UNICEF (India Country Office)
    73, Lodi Estate, New Delhi – 110003, India
    (Non-formal education, water and sanitation, health, children)
46. THE WORLD BANK
    55, Lodi Estate, New Delhi-110003, India
    (Funds)
47. WORLD HEALTH ORGANISATION (WHO)
    Indraprastha Estate, New Delhi-110002, India
    (Health)
48. WWF INDIA
    172 B, Lodi Estate, New Delhi-110002, India
    (Conservation of Wildife)
49. MINISTRAY OF ENVIRONMENT AND FORESTS
    Paryavaran Bhavan, CGO Complex, Phase II,
    Lodi Road, New Delhi-110003, India
    (Funds, Ecoclubs of Protection of Enviornment)
50. COUNCIL FOR ADVANCEMENT OF PEOPLES'S ACTION AND RURAL TECHNOLOGY (CAPART)
    Gurunanak Foundation Buildings, New Mehrouli Road, New Delhi-1100067, India
    (Rural Education and Development)
51. THE FORD FOUNDATION
    55, Lodi Estate, New Delhi-3, India
    (Finance for scientists attending International Conferences, Seminar and Workshops)
52. DEPARTMENT OF SCIENCE &TECHNOLOGY
    Technology Bhawan, New Mehrouli Road, Near I.I.T.,
    Hauz Khas, New Delhi -110016, India
    (Financial support for projects)

53. CENTRAL POLLUTION CONTROL BOARD (CPCB)
Parivesh Bhavan, East Arjun Nagar, Shahadara, New Delhi-110032. Phone : 2217213

54. TATA ENERGY RESEARCH INSTITUTE (TERI)
Darbari Seth Block, Babitat place, Lodi Road, New Delhi – 110003. Phone: 4622246, 4601550

55. SALIM ALI SCHOOL OF ECOLOGY
University of Pondicherry Uducherry-605014, TN

56. PEOPLE OF FOR ANIMALS (PEA)
A4, Maharani Bagh, New Delhi-110065
(An organisation of the noted environmentalist Ms. Maneka Gandhi, M.P.)

57. CREED (Center for Research on Environment, Ecology and Development)
3rd Floor, JP Howa, 118, Shapur Jat, Near Asiad Games Village, New Delhi–110019

58. INDIA SOCIETY FOR TECHNICAL EDUCATION (ISTE)
I.I.T. Campus, Hauz Khas, New Delhi – 110076
(Funds Winter Schools/Workshops/Seminar's etc. and establishing Environmental Citizenship Cell, publishes a book giving the addresses of all Private and Government Polytechnics and Engineering colleges of India)

59. BIS (BUREAU OF INDIAN STANDARDS)
Manak Bhavan, 9, Bahadur Shah Zafar Marg, New Delhi-2
Phone : 3230131, 3375, 9402 for IS Codes

60. J.N. TATA ENDOWMENT LOAN SCHOLARSHIPS
Bombay House, 24, Homi Mody Street, Mumbai-400 001

61. BRITISH HIGH COMMISSION
Chanakyapuri, New Delhi-110 021, Ph: 601371

62. COUNCIL OF SCIENTIFIC & INDUSRIAL RESEARCH
CSIR Complex, NPL Campus Pusa, New Delhi-110012

63. US AID MISSION
28, Institutional Area, Gulab Hotel Road, New Delhi-16

64. USA ACADEMY (UNIVERSITY PLACEMENT SERVICE)
777, South State Road, Suite 5X, Fort Launderdale – 33068 Florida, USA Ph: (954) 9689312 Fax : (954) 344 2425
E-mail: us degree@aol.com

65. UNDP DIVISION OF INFORMATION
Room DC-1-1900, 1-UN Plaza New York, NY-10017, USA
66. PROJECT OFFICER, COMMISSION SERVICE COUNCIL
Marl Borough House, Pall Hall, London Swit SHX, UK
67. CHIEF OF FELLOWSHIPS SERVICE
UN Dept. of Tech. Cooperation for Development (UNDTCD)
1-UN Plaza, New York, NY-10017, USA
68. INDIAN ACADEMY OF SCIENCES
C.V. Raman Avenue, P.B. 8005, Bangalore, 560 080
Ph. : 080-2334 2546, 2334 2943
(Releases 'RESONANCE'–A Journal of Science Education)
69. SOCIETY FOR ENVIRONMENTAL COMMUNICATIONS/CENTER FOR SCIENCE AND ENVIRONMENT
41, Tughlakabad Institutional Area, New Delhi-110 062
(Releases the most popular Fornightly "Down to Earth")
70. INDIAN SCIENCE CONGRESS ASSOCIATION (ISCA)
14, Dr. Biresh Guha Street, Calcutta-700 017.
(Releases 'Everyman's Science Quarterly Bulletin)
71. NEERI (National Environmental Engineering Research Institute)
Nehru Marg, Nagpur, India-440 020
(Releases "Environmental Health" – a Research Journal)
72. NATIONAL INSTITUTE OF ECOLOGY AND ENVIRONMENT (NIEE)
C/o, Environmental Protection Society,
Post Box 9, Rajeshwari Nagar, Kakinada, A.P., India
Ph. 0884-23525461/2378313, 98491-36135
Email: director@natureindia.org
73. ENVIRONMENTAL RESEARCH ACADEMY (EnRA)
58, Siva Sadan, Seetammadhara North Extn.
Visakhapatnam 530013, A.P. India
Cell: 09490 26 5345
E-mail: ratna_madireddi@yahoo.com
(Releases "Environ" an yearly Bulletin)
Please Contact: Mrs. M. Ratna kumari, M.A.,

for published books on Wildlife in India.
Environment, Biodiversity, Membership in
EnRA; Organisation of Seminars,
Projects Preparation, Guidance for Ph.D.,
Funds from Agencies, EIA, Pollution etc.
Phone: 91-891-2784100
Fax : 91-891-2560479
Mobile : 91-09490265345
President : Prof. M.V. Subba Rao, Ph.D.,

Courtesy: EnRA, Visakhapatnam

UNIT-I

1
Chapter

INTRODUCTION

ORIGIN AND NATURE OF SOILS

The United States Department of Agriculture (1951) defined soils as "the collection of natural bodies occupying portions of the earth's surface that support plants and that have properties due to the integrated effects of climate and living matter acting upon parent material, as conditioned by relief, over periods of time."

Of many different kinds of soil on earth, each has its own special collective characteristics upon which its behaviour depends. It is necessary to distinguish between soil as an important substance and soil as a natural body.

The earth's crust is the origin of the soil. During the process of weathering, primary rocks produce what is known as 'soil parent material' and the rocks from which this material is produced called 'parent rocks'. Soil parent materials are now recognized in four main classes :

I. Those formed *in situ* from soft rocks such as chalk and volcanically.

II. Those formed *in situ* from hard rocks which may be igneous (*e.g.*, granite and basalt), sedimentary (*e.g.*, limestone and sandstone) or metamorphic (*e.g.*, gneiss and marble).

III. Those transported from their place of origin before weathering to form soils. This is the most important of the parent material classes and includes all material moved by water (such as alluvium, colluvium, lacrustine deposits and beach sediments), by wind (such as loose and eolian sands), and by glacial phenomena (such as glacial drift and tilt).

IV. Those formed from organic material such as peat.

Rocks are weathered mechanically by the action of frost, water, wind and chemically by such processes as oxidation, hydration, hydrolysis and solution in Carbonic acid. Similar or identical rocks may produce quite different materials basing on the kind of weathering.

These products undergo further change to produce soil, as we know it. This process does not necessarily occur at the same time or place as production of parent material as the soil forming processes may have proceeded concomitantly with rock weathering. In many soils as now found, more than one parent material may have been involved in their formation.

ROCKS (MATERIALS OF THE EARTH'S CRUST)

The outer part of the earth is the earth's crust. The most significant outer crust of the earth is part of lithosphere. Down to a depth of 16 km from the surface of land, 95% of the earth materials found in the earth's crust, consists of rocks. Rocks are made up of minerals which are naturally occurring solid materials.

Minerals in Rocks

Minerals are found mostly in solid state. Each mineral usually contains two or more simple substances called elements of which the whole earth is made. Out of about 2000 different minerals, only 12 are common to all of our earth. These 12 minerals are called the rock formers. Common minerals are formed of eight most abundant elements in nature. Out of these, silicates, carbonates and oxides make up a large group of them. About 87% of minerals in the earth's crust are silicates. Quartz has two elements, silicon and oxygen, united together to form carbonate of lime. Minerals are chemical substances found in nature and may be either elements or compounds. There are metallic minerals like copper, lead and gold as well as nonmetallic ones like mica, gypsurn and quartz.

Types of Rocks

Rocks are classified on the basis of mode of their formation. These are igneous, sedimentary and metamorphic rocks.

Igneous rocks are formed by solidification from a liquid state. These are formed by gradual cooling of molten rock called magma, chief source of metal ores which gets erupted by volcanoes. As the earth cooled to a solid crust the original crust of the earth had igneous rocks. Igneous rocks are therefore, known as primary rocks. Basalt and Granite are examples of igneous rocks.

Sedimentary rocks are made up of sediments deposited usually on the floor of seas and lakes. Sediments consists of particles of gravel, sand, silt or clay. The loose sediments get compacted into rocks by pressure of overlying sediments and presence of cementing materials like lirne. Limestone and sandstone are sedimentary rocks (also called stratified rocks).

Metamorphic rocks are those formed by change of form of pre-existing igneous or sedimentary rocks, when they were subjected to extreme heat or pressure or both. Limestone gets converted to marble, a metamorphic rock. Sandstone gets metamorphosed to Quartzite.

Rock Cycle

All rock materials originating from beneath the earth's surface form igneous rocks. Due to exposure to weather, igneous rocks are eroded by various agents. This eroded material changes into sedimentary rocks at some places and at some points of time. Either of the two rocks are likely to change into metamorphic rocks in course of time. A change into sedimentary rock rakes place at the earth's surface and conversion into metamorphic rock takes place within the earth's crust. The sedimentary rock may again be buried so deep that they melt forming igneous rocks. This change of one rock type into another type under different conditions is known as rock cycle, shown below. Since matter of earth's crust is not lost the process becomes cyclic.

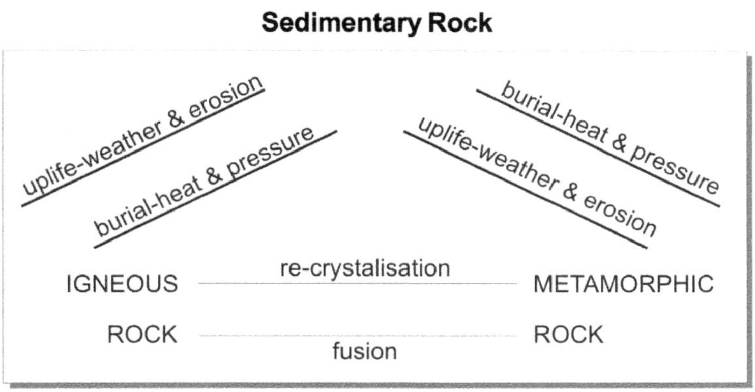

For a better understanding, the process of soil formation from parent material can be divided into two phases. The first phase involves addition and admixture of organic matter and the other involves solution and translocation of certain constituents, which is known as 'aluviation'. Aluviation itself can be further divided into mechanical aluviation in which finer particles, usually move downwards by washing and chemical aluviation which involves the partial decomposition of colloidal matter with subsequent downward movement of the decomposition products.

During or after the soil formation, unrelated effects such as volcanic action, floods, erosion and human activities can modify its nature.

The eventual soil is chemically, physically and biologically a complex dynamic system, the constituents of which are constantly undergoing change. Hall (1949) likens soil to a three phase system of solid, liquid and gaseous components with constantly shifting equilibrium.

For further understanding and discussion on the origin of soils, it is recommended to refer to the standard works on soil genesis and constitution (Hall, 1949; Russell, 1961). From the analyst's point of view, it is apparent that soil, even as material in a bottle, provides a stimulating challenge to its soil capabilities.

Table 1: Different layers of earth's interior

Name of the layer	*Chemical composition*	*Average thickness (km)*	*Density*
1. Lithosphere			
(i) Outer crustal layer (solid)	SIAL	8 to 45	
(ii) Inner layers (party molten)	or outer silicate layer, rich silica and aluminium	50 to 90	2.75 to 2.90
2. Mantle			
(i) Inner silicate layer	SIMA (in part)	100 to 400	3.1 to 4.75
(ii) Transitional zone	SIMA (wholly)	2400 to 2700	4.75 to 5.0
3. Core			
(i) Outer metallic	NIFE	2000	5.1 to 13.0
(ii) Inner metallic	Barysphere (heavy metallic rock)	1500	

2
Chapter

LITHOSPHERE

The word "Lithosphere" refers to the layers of rock material on the earth's surface, both on the continents and ocean floors. Lithosphere forms a relatively thin (50-100 km thick) crust, which is thicker in the contents that on the ocean floors. The outer, crustal layer is of lighter density than the inner layers. Since the outer crustal layer comprises of rocks rich in silica and aluminium, it is called the SIAL is solid and 8 to 45 km thick. Inner layers are 45 to 100 km thick and in partly molten state. Below the lithosphere lies the mantle, which extends upto a depth of 2900 km. Mantle, like lithosphere can also be divided into:

(I) Outer, 100 - 400 km thick, asthenosphere, (i.e. inner silicate layer and outer silicate layers that form inner part of lithosphere) having materials rich in silica and magnesium, hence called SIMA layer.

(II) Inner, 2400 - 2750 km thick, mesosphere which is transitional zone of mixed materials and silicates; also known as wholly SIMA.

The central part of the earth core is about 3500 km thick, its outer part, about 2000 km and the inner part about 1500 km thick. The core consists of nickel and iron, hence also called NIFE, and accounts for the earth's magnesium. (Fig. 2.1).

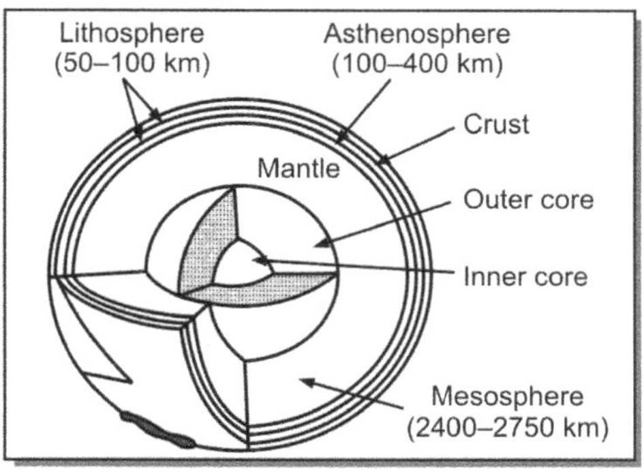

Fig. 2.1: *A Section through the earth showing its structure*

LAND RESOURCES

Only about 10% of earth's land area is currently under agricultural production, (*i.e.* out of total of 144.8 million Sq. Km is 14.78 million Sq. Km). Perhaps four times as much land could potentially be converted for crops, but much of this land serves as a refuge for cultural or biological diversity or suffers, from constraints, such as steep slopes, shallow soils, poor drainage, tillage problems, low nutrient levels, metal toxicity or excess soluble salts or acidity, that limits the types of crops that can be grown there.

In India, our land is 329 million hectares. Out of which 51% is agricultural purpose. The barren land is 21 Mha. The available land for potential land use is 266 Mha, 23 Mha is fallow land, 13 Mha is pasture land and 30 Mha is tree cover.

The current world average land availability of 0.26 ha (0-64 ac) of crop land per person will decline to 0.17 ha (0.42 ac) by the end of year, 2025. In Asia, including India, the crop land will be even fall, *i.e.* 0.09 ha (0.22 ac) per person due to increasing human population.

In the developed countries, 95% of recent contribution of growth in agriculture is due to the improved crop varieties or increased fertilization, irrigation and pesticide use, rather than from bringing new land into cultivation. However, there would be serious ecological trade off's in putting much of this land into agricultural production. In many cases clearing land for agriculture in the tropics has resulted in tragic losses of biodiversity and the valuable ecological services that it provides. Ultimately, much of this land is turned into useless scrub or semidesert. On the other hand, there are large areas of rich subtropical grassland and forest that are well watered, have good soil and could become productive farmland without unduly reducing the world's biological diversity.

LAND FORMS

Generally the surface of the land is not uniform in height or appearance in different areas or places. On the continents, the mountains, plateaus and plains are major land forms. Due to sudden or gradual changes caused by earthquakes, volcanoes, floods, depressions, Tsunami etc. the land forms develop.

LAND FORMATION

Due to the action of two natural processes on the surface of land, the Landforms are formed.

1. External processes that take place in the atmosphere and hydrosphere and affect the land surface. These act slowly from high land to low lands and deposits materials.
2. Internal processes that take place in the interior of the earth and cause changes on the land surface. These processes cause movements of the earth's crust leading to formation of mountains and plateaus.

The type of rocks has an influence on the land forms developed in an area. Strong rocks like granite, quartzite and limestone are not easily weathered. They give rise to hills and mountains. The soft rocks like clay, shale, loose sand forms a bluffs of knolls but turn into plains and valleys, sedimentary rocks forms plains or slopes. Climate of the area has role in the formation of land forms.

TYPES OF LAND FORMS

On the basis of the shape, there are three major land forms - mountains, plateau and plains.

IMPORTANCE OF LAND FORMS

Each land form has its own importance to man. Mountains have protected mankind since long time. Man's habitation is isolated and restricted to his movements. Mountains provide natural barriers between nations. Agriculture is scanty on mountains. Valleys between mountains are used for cultivation. There is a developed transport system between plains and mountains.

Forests are important natural resources. There are not only places of tourists but also sources of natural wealth to local people. Some plateaus are rich in minerals and gems. Particularly in tropical and temperate regions, the plains are very fertile for cultivation, easy mode of communication, transport and other facilities. These are areas of rich agriculture and industrial growth. Hence people are interested to live in plains rather than mountains.

USE OF THE LAND

Land, which is a nature gift, is useful for production of variety of crops and is useful of livelihood for both men and other organisms living on the earth. To cater the needs of the increasing population of India, the land is used for food production. The land is also used for growing fruit crops and vegetables, which are used in day to day life of the human beings.

To grow the plants and to uptake of the nutrients, which are essential for plant growth and to give more yield. Land is the main base. It is also useful of all living animals, creatures and micro-organisms, plants, etc. for shelter and home.

Productivity: The increased productivity is much more essential for the entire population who depends upon plants. The percentage of production is being increased by using mechanized agricultural practices, increased use of fertilizers, need based application pesticides. Productivity is also essential for growth and development of animals that depend on various plants and parts of the plant.

MANAGEMENT OF LAND USE

Today increasing use of land due to rapid industrialization, exploding the population and deforestation has aggregated the soil pollution problems. Negligence use of land can damage soil, which result in soil infertility, soil erosion, desertification and degradation of watersheds and catchments. The entire land is under stress due to urban-industrial and agricultural sprawl. Our croplands are losing fast fertile soil because of uncontrolled confusing statistics. For proper land use, planning, authentic figures should be envisaged by the concerned departments (Agriculture, Forestry, Revenue and records etc.) Today India has the lowest man-land ratio, barely 0.48 hectare per capita. Japan and France have planned a most effective scheme of land preservation. Researchers are still going on to develop the best strategy of land use.

Following steps can be beneficial to Land use, they are :

1. Preparation of accurate land use data through Remote sensing.
2. A nationwide survey programme to set up land use policy.
3. Preparation of management plans for land amelioration.
4. Reviewing existing legislation's and to amend them for proper use of land.
5. Educating the people by holding land use classes.
6. Systematically controlling population growth.

   In addition to the above micro and macro level activities should be practiced systematically.

MICRO LEVEL ACTIVITIES

1. Farmers must be educated on the right and beneficial agricultural methods to be adopted.
2. Local agricultural extension workers need support and training for the benefit of the farmers.
3. A village can be adopted through NGOs to takeup a community development project as a pilot project.
4. Saplings may be planted and maintained by the villagers or NGOs. Commercial tree or useful hybrid saplings make greater incentives to the villagers and when handed over to women who can be motivated to involve themselves in tree cultivation, not only to benefit but also afforestation for sustainable development of land use to them.

MACRO LEVEL ACTIVITIES

1. Collective participation in water and land management programmes which are :
   - Tree planting on a large scale on barren lands.
   - Contour bunding and utilize khas khas grass.
   - Construction of small water reservoirs and soak pits, trenches etc.
   - Controlled grazing lands.
   - Distilling of tanks and reservoirs.
2. Influencing the Government on major issues such as the location of heavy industries that would displace or disrupt the lives of the people.
3. Development of rural areas.

> *"Today, the environment is at risk. Unless we can come to terms and unless we can make peace with the Environment, we are not going to survive".*
>
> **Dr. NOOL BROWN,**
>
> Spl. Representative to the UNEP, UNO

3

Chapter

Definition of Soil

Soil is a marvellous substance, a living resource of astonishing beauty, complexity and frailty. It is a complex mixture of weathered mineral materials from rocks, partially decomposed organic molecules and a host of living organisms. It can be considered as an ecosystem by itself. Soil is an essential component of the biosphere. But it must be nurtured and cultivated to bring it to its highest potential.

Any part of earth's surface bears a covering of vegetation. Soil is thus usually defined as "any part of earth's crust in which plants carry roots". Muddy bottoms of ponds, porous rock surfaces, ravines or glacial deposits, bottoms of lake etc., all are the soils. But this is a limited definition of soil, as we know that soil is actually formed as a result of long term process of complex interactions leading to the production of a mineral matrix in close association with interstitial organic matter takes fairly a long time. Such a modification is actually the result of interactions between climatic, topographic and biological effects.

Soil is thus not merely a group of mineral particles. It has a biological system of living organisms as well as some other components. It is thus preferred to call it as a soil complex.

Generally soil is a thin layer of material that lies over rocks, covering most of the land. It may be few continents deep, or it may extend several meters down to the rock below. Soil forms the link between the abiotic and biotic parts of a terrestrial ecosystem. Plant roots go through soil and take in water, minerals and oxygen. It has four main components of mineral particles (upto 60%), organic matter (about 10%), water (upto 35%) and air (upto 25%). The mineral particles of soil area derived from underlying rock as it undergoes weathering.

4
Chapter

Formation of Soil

The five factors responsible for soil formation, they are :

1. Parent Material
2. Climate
3. Life
4. Topography
5. Time

PARENT MATERIAL

Nature deposited sandstone in some places and limestone or shale in other places. At still different locations, molten rock (magma) from the interior of the earth was slowly pushed upward, forming and exposing granite. But whatever bedrock, nature left, weathering by rainwater, heating and cooling, freezing and thawing. The action of plants and animals have been working for centuries to wear the rock away. At the same time that rocks have been rotting, other forces in nature have been building a soil from the decaying rock pieces. New chemicals have been formed in the soil from the decay products of the rocks and plants. Humus has been built from plant and animal residues. Clay minerals have been synthesized into something both physically and chemically different from the minerals in the parent rocks. They are montmorillonite and kaolinite. Rocks are mixtures of minerals. The soils developed from sandstone are coarse textured and sandy soils. Soils from shale are clay in texture but are not very fertile. Soils from granite bedrocks are usually sandy loams of low fertility. In contrast, the soils developed from limestone bedrock are usually dark in colour, fine in texture and very fertile.

CLIMATE

Changes in temperature from day to night and from season to season would break the rock into scale like pieces. Rainwater would dissolve some of the limestone and often other parts of rock wetting and drying, heating and cooling would cause expansion and contraction of the loosened rock particles. In some of the tiny cracks and depressions there would be enough water and plant nutrients to support lower forms of plant life. Thus soil is formed. Depending upon the climate, from limestone bedrock, the number of years of weathering and synthesis of minerals are required to make soil.

LIFE

Lower forms of life such as lichens are the first to grow under the harsh environments that exist on slightly weathered limestone bedrock. After many years of growth and death of these tiny plants, the environment is thereby improved, mosses become established. Generation after generation of lichens and mosses grow, reproduce and die until the organic matter from these decaying tissues helps to make a better place for seed bearing higher plants to start growing. Many annual weeds will migrate into the area followed by grasses, shrubs and finally trees if the rainfall is adequate. If rainfall is less, the grasses will be the dominant vegetation. During soil formation, bacteria and fungi and many birds and other animals help the process of disintegration and decomposition of rocks into soil. Earthworms, ants, arachnids, millipedes, centipedes and many rodents are always manipulating the soil material and thus continuing the formation of soils from rocks. There are about 1,800 species of earthworms throughout the world. The common earthworm (night crawlers) belongs to the family Lumbricoid. All species eat coarse organic matter and burrow into soil and hence the soil is both chemically and physically a better medium for plant growth. The holes of old earthworms are ideal channels for roots of plants. Hence they are good friends of farmers.

TOPOGRAPHY

The large boulders having small depressions in the center in order to hold more rainwater (without the slope formation of the rock), the process of soil formation, plant growth and plant succession would be faster. In the hill side areas, the soil formation is slow, and the more level areas of rocks, the formation of the soils is faster. Too much water can retard soil development.

TIME

A soil formed from parent material by continuous weathering and the accumulation of organic matter on the surface. The process of change of the parent material changes to soil is always difficult to determine. But as lime and other salts, clay, iron and humus compounds move downwards as parent material weathers. The organic matter accumulates on the surface, the soil is being formed. None of them can be seen or measured. There are two aspects of the age of a soil - absolute and relative age. Absolute age refers to the number of years, the soil has been forming (developing), without interruption from parent material. Relative age refers to the stage of maturity of the soil as indicated by the soil profile. The kind of soil that has developed from a particular parent material in a designated climate with certain plants and animals under a given topography over a designated time will have a physical, chemical and biological character.

5
Chapter

KINDS (TYPES) OF SOIL IN INDIA

There are four major types of soils in India (Fig. 5.1).

1. Indo Gangetic Alluvium
2. Black Cotton Soils
3. Red Soils
4. Laterite Soils

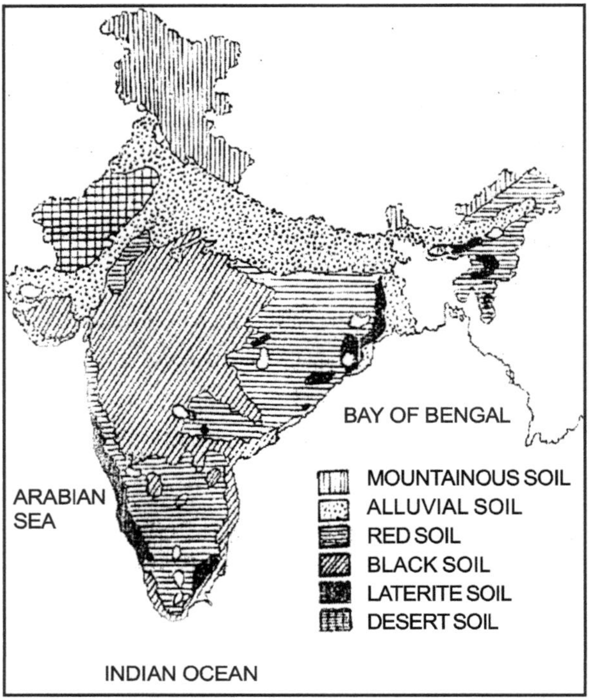

Fig. 5.1 : *India–Major soil types*

According to ICAR, the soils are divided into 9 categories.

1. Alluvial Soils
2. Desert Soils
3. Saline and Alkaline Soils
4. Peaty and Marshy Soils
5. Foot Hill or Terrain Soils
6. Mountain and Hill Soils
7. Black Soils
8. Red Soils
9. Laterite Soils

1. ALLUVIAL SOILS

These are mostly from clayey to sandy loam in texture and are generally acidic in nature. These soils are deficit in lime, Phosphoric acid and humus. The soils are red colored and differ in different parts of the country. Although the soils are rich in potash, Phosphoric acid, lime and organic matter. They are deficient in Nitrogen and humus contents.

2. DESERT SOILS

They occur under arid and semi arid conditions and occupies large tracks in Rajasthan, Haryana and South Punjab. These soils contain high percentage of soluble salts, low loss on ignition and poor in organic matter. These soils may be reclaimed if proper facilities are available.

3. SALINE AND ALKALINE SOILS

These soils occur on about 68,000 km^2 of area of Bihar, U.P., Haryana, Punjab, Rajasthan and Maharashtra. Soils give rise to saline and alkaline. The Alkaline soils are deficite in Calcium and Nitrogen and are highly impervious and have very low water holding capacity. On these soils, a wide variety of crops including rice, wheat, cotton, banana, sugarcane and coconut grows.

4. PEATY AND MARSHY SOILS

These soils are generally submerged under water during the monsoon. These soils are black, heavy and highly acidic contain 10 to 40% of organic matter. Marshy soils are found in coastal tracks, especially in Bhitarkanika of Orissa and Sundarbans of West Bengal, which contain 18% of organic matter.

5. FOOT HILL OR TERRAIN SOILS

Particularly deficient in Phosphate, rich in Nitrogen and organic matter. These soils are mainly sandy.

6. MOUNTAIN AND HILL SOILS

Rich in humus and very fertile for cultivation of plantation crops like tea and temperate fruits and may be used for growing paddy.

7. BLACK SOILS

These soils are found in the states of Andhra Pradesh, Tamilnadu, Karnataka, Madhya Pradesh and Maharashtra. They are rich in fertility, rich in lime, high $MgCO_3$ and contain sufficient quantities of potash and Phosphoric acid. These black soils are unsuitable for heavy irrigation and good for crops of cotton.

8. RED SOILS

The red soils differ greatly in colour, depth and fertility. Since these soils are airy, hence they are useful for cultivation.

9. LATERITE SOILS.

These soils are generally poor in Nitrogen, Potash, Potassium and organic matter. They produce good crops.

Soil families are groups of soils having similar percentage of sand, silt and clay, Similar kinds of minerals and temperature regimes and other similar properties required for the production of principal crop plants.

To increase the efficiency of food production and in anticipation of central and state land use legislation - soil conservation service (Agriculture Department) started a system of using soil maps to locate prime and unique agriculture lands. In India the total land area is 239 million hectares.

6

Chapter

SOIL PROFILE

A vertical section (profile) of a soil usually shows three natural layers known as "Horizons". The horizons can be identified by colour changes, texture, structure or in chemical compositions.

The top soil or surface mineral layer is the A-Horizon, the subsoil is the B-Horizon, the parent material (bedrock) is the C-Horizon and the underlying bedrock is R-Horizon. A soil profile with a third dimension is known as a pedon. The major horizons of mineral soils are as follows (few soils have all of these horizons) (Fig. 6.1).

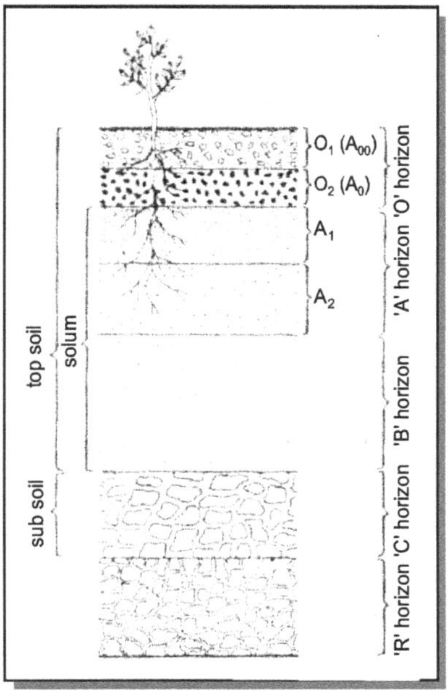

Fig. 6.1 : *Hypothetical diagram of the soil profile to show principal horizons*

The different horizons of a soil profile historically in Russian terminology, were classified into ABC terminology. However, according to the present one in general use, the soil profile (Fig. 6.2) consists of the following five main horizons.

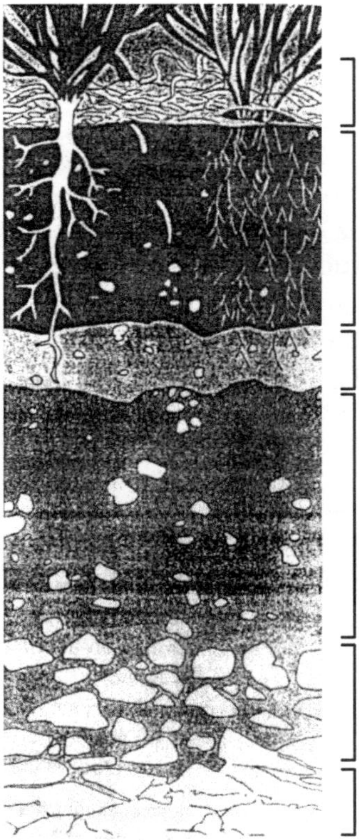

Fig. 6.2 : *Soil Profile showing possible soil Horizons*

O-HORIZON

An organic layer of fresh and decaying plant residue at the surface of a mineral soil.

A-HORIZON

The mineral horizon at or near the surface in which an accumulation of humidified organic matter is mixed with the mineral matter. Also a plowed surface horizon.

E-HORIZON

The mineral horizon in which the main feature is loss by leaching of silicate clay, humus, iron or aluminum or some combinations of these.

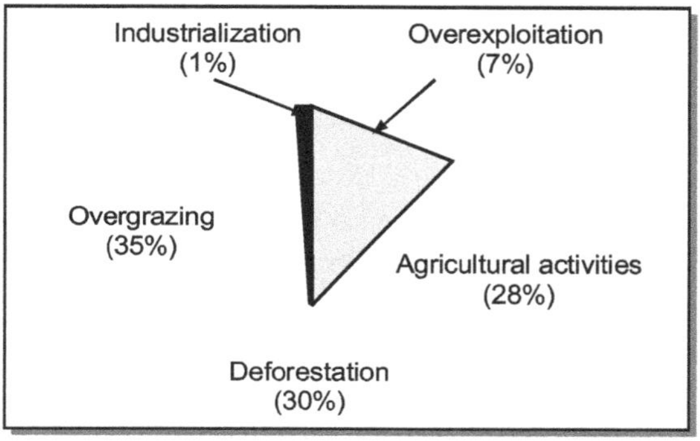

Fig. 6.3 : *Soil Degradation world wide for all land uses*

B-HORIZON

The mineral horizon below O, A or E Horizons. The B-Horizon is in part a layer of transition from the lying horizon to the underlying C-Horizon. The B-Horizon has an accumulation of clay, sesquioxides, humus or combination of these. Prismatic or blacky structure, red or brown colour or combination of these. The combined A and B horizons are generally called the solum or true soil. If a soil does not have a B-Horizon, the A-Horizon alone is the solum.

C-HORIZON

The mineral horizon or layer, excluding indurated bedrock, that is little affected by soil forming processes and does not have the properties typical of the A or B Horizon. The material of a C-Horizon may be either like or unlike that in which the solum formed. If the material is known to differ from that in the solum, the arabic number 2 proceeds the latter C (*i.e.* C2).

R-HORIZON

Consolidated rock (unweathered bedrock) beneath the soil, the rock commonly underlies a C-Horizon, but it can be directly below A or B Horizon if there is no C-Horizon.

7
Chapter

CLASSIFICATION OF SOILS

A system of soil classification is known as Soil Taxonomy. The soil classification system includes order, suborder, great group, family and series. The nomenclature of the system above the level of the soil series is new. The names have been systematically coined mostly from Latin and Greek origin.

The six categories of soils in USA consists of Orders 10, Suborders - 47, Great groups - 200, Subgroups -1000, Families - 2000 and Species: 12,620. (US system of Taxonomy, 1984). In 1989, an eleventh soil order was officially recognized (Order 11) is predominantly found in New Zealand.

Now, there are 11 soil orders. They are :

1. ENTISOILS (NEW)

Very young soils with little or no horizon development. (8% in the world).

2. VERISOILS (VERTO = TURN)

Self-swallowing soils that have large cracks during the dry season into which surface soil falls. They contain much clay that swells when wet and causes a "churning" process (2% in the world).

3. INCEPTISOILS (INSEPTUM = BEGINNING)

Young soils that have had time to develop only the more rapidly formed horizons such as histic, umbric or ochric epipedons (surface layers) or cambric horizons (9% in the world).

4. ARIDSOILS (ARID = DRY)

Soils of dry regions (19% in the world).

5. MOLLISOILS (MOLLIS = SOFT)

Soils with mollic epipedons. Wet soils and soils with argillic horizons are included if they have mollic epipedons (9% in the world).

6. SPODOSOILS (SPODOS = WOOD ASHES)

Refers to light gray (A2 -Horizon) - soils with spodic horizons; They are highly leached soils of cool, humid, forested areas (4% in the world).

7. ALFISOILS (AL, FE = ALUMINIUM, IRON)

Soils that have argillic horizons and are not too highly leached. The epipedon may be orchric or umbric but not mollic (13% in the world).

8. ULTISOILS (ULTIMATE)

Similar to Alfisoils but more leached. So that they have more red and yellow (6% in the world).

9. OXISOILS (OXIDES)

Soils with an oxic horizon, which form in an old, humid, tropical environment (8% in the world).

10. HISTOSOILS (HISTOS = TISSUES)

Organic soils (1% in the world).

11. ANDISOLS (ANOS = BLACK)

Black mineral soils with a melanic epipedon and 60% of entire soil volcanic ejecta, including 5% volcanic glass (soils in New Zealand).

Main Environmental Problems are:

- *Land degradation and Soil erosion.*
- *Ecological disturbances - Soil Ecology.*

- *Loss of Forest and Vegetation - Deforestation.*
- *Recurring floods.*
- *Interruptions and disturbances to drainage systems.*
- *Impairment of valuable fisheries - Aquatic ecology. Aesthetic degradation.*
- *Air, Water, Soil and Noise pollution*
- *Accidental spill of Hazardous materials.*
- *Desertification*

8
Chapter

Characteristics of Different Soil Types

SAND, LOAMY SAND AND SANDY LOAM

These are well drained and aerated and workable for most of the year. They are very light to handle and quick to warmup in spring. Unless they have a very high organic matter content, they are prone to drying out very quickly and additional watering will be needed. This extra watering will also help to wash out the plant nutrients and lime from the soil, so they are likely to be acidic (except for some coastal soils). They are often referred to as "hungry soils" and need lots of extra feeding. With careful management, however, they can be amongst the most productive soil types. Medium loam, Sandy clay loam, Silt loam. These are the average of soil types. They achieve a good balance between the ability to be very productive and the minimum of attention. The medium loam group is probably the best in this regard. (Fig. 8.1)

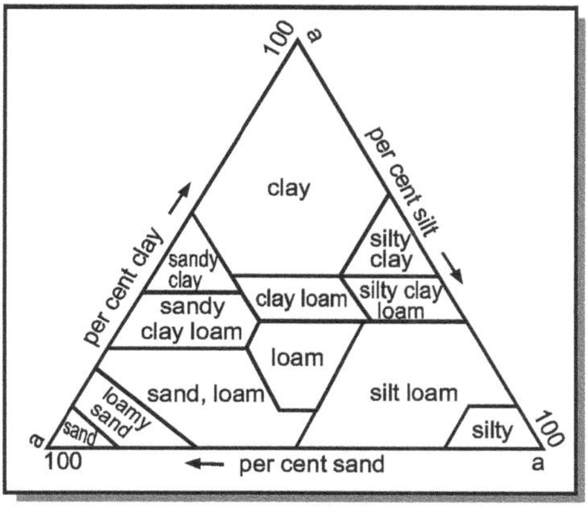

Fig. 8.1 : *Diagram to show classification of soil texture*

Clay, Sandy Clay, Clay Loam, Silty Clay Loam, Silty Clay and Silt Soils

Although these soils are difficult to work and manage, they usually have good supplies of plant nutrients and lime. The main drawbacks are the high water holding capacity and the effort required to work them. They will need to catch just the right weather conditions to avoid hard work and damage to the soil structure. The use of heavy machinery (especially rotavators) should be avoided at all costs, particularly when the soil is wet.

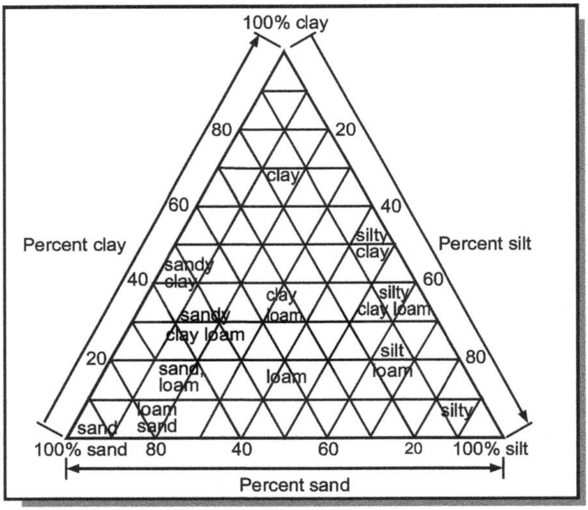

Fig. 8.2 : *Soil Texture*

Soil texture is determined by the percentage of clay, silt, and sand particles in the soil (Fig. 8.2). Soil with the best texture for most crops are loams, which have enough large particles (Sand) to be loose, yet enough smaller particles (Silt and clay) to retain water and dissolved mineral nutrients.

Peat Moss or Fen Soils

They are provided not too acidic and have effective sub drainage. These are probably the best natural soils available. They are rich in plant nutrients, and are easily workable. It is possible to convert existing soil into peat type soil by adding large amounts of organic matter. Some of the keenest exhibition growers do just this. It can be time consuming and expensive at first, but once they get there life becomes much earier. It must avoid making soil too acidic though and careful choice of organic matter is needed.

Chalk Soils and Limestone Soils

These are the soils that contain a high proportion of chalk or lime. In fact it overrides their normal particle size classification. They are often very shallow soils and severely limit the types of plants that can be grown successfully in them. If a person have a soil of this type and are not happy with the range of plants, it will allow to grow, probably the best thing can do is to move a new area and check the soil first. If you cannot move, the most sensible course of action is to limit yourself to the plants that will grow in chalky soils. Trying to change the soil is usually a difficult task and quite expensive.

Hydrogen Ion Concentration (pH)

pH is the measurement of the acidity or alkalinity of soil. The amount of lime (Calcium) contained in soil and the type of soil determine it. In general, British soils range from about 4 to 8 (the word pH is a boffins term, it stands for "potential Hydrogen ions"), as well as affecting soil acidity. It is worth remembering that lime is the fourth most important plant nutrient. Some plants are very sensitive to soil pH probably the best known examples of this are (the majority of) Rhododendrons and heathier's which will not tolerate lime in the soil. On the other hand, Clematis and Daphnem, are prefers lime soil.

It is not generally known howerver that many plants are sensitive to pH levels. The plant pH preference lists will help you to see which plants prefer to grow at particular pH levels. For many plants a deviation of 0.5 is permissable.

pH Affects Fertilizers and Plant Foods

Any plant that is suffering from a deficiency of lime (i.e. growing in a pH group that is too low) will show stunting symptoms and the growing tips of the plant are yellow and deformed, whilst the lower parts of the plant remain unaffected. This is often accompanied by short and stubby root growth instead of long fibrous ones. At very low pH levels, acids escape into the soil and dissolve Aluminum and Manganese. These are poisonous to some plants.

On the other hand, plants growing in a alkaline soil (where pH is too high) can also be adversely affected. This is generally more because of secondary affect on other plant foods than because of an excess of Calcium.

Classification of Indian Soils

Indian Soils	Distribution
Red Soils	
Red Loam Soils	Madhya Pradesh to Orissa
Red Sandy Soils	Tamil Nadu, Karnataka, Andhra Pradesh
Laterite Soils	Southern India, Assam, Orissa, Madhya Pradesh, Maharashtra, Gujarat
Red and Yellow Soils	Madhya Pradesh, Orissa

Soil Conservation

The aim of Soil Conservation is to protect the soil from erosion and maintain productive capacity of soil. Soil erosion is a natural phenomenon. Several methods are employed to check erosion. Vast lands are rendered useless due to industiralisation and urbanisation. Erosion has posed a serious problem in various countries. In India, the desert of Rajasthan is a good example. Several programmes like FAO, UNDP were made to strengthen (All India soil and landuse survey). A programme for restoration of fallow lands was launched in eight states: Andhra Pradesh, Tamil Nadu, Bihar, Karnataka, Madhya Pradesh, Maharashtra, Rajasthan and Uttar Pradesh. But the problem of soil erosion continues.

There are several methods for conservation of soils:

Biological Methods: It is helpful in checking the soil erosion. The methods are:

(*a*) Agronomic practices which include contour farming, tillage and keeping the land fallow, crop rotation, mulching, strip cropping etc.

(*b*) Agrostological methods which include cultivation of grasses, retaining in land, Afforestation and reforestation, checking of over grazing.

(*c*) Dry farming is useful in areas with very low rainfall and uneven distribution of rainfall. In such areas mechanical methods are employed like subsoiling, pan breaking, basin leeching, contour terracing, terrance outlets, gully contours etc.

International Conference on Soil Conservation

***Consoil, 1997:** International Conference on Urban Ecology-UFZ Center for Environment, Leipzig, Germany, 25-29 June 1997 (Author participated and presented a paper on "Land use as a controlling factor in terms of Urban Ecology in India" (Received the Appreciation award from the Organisers).*

9

Chapter

PHYSICAL CHARACTERS OF SOIL

The physical characters of a soil and its suitability for plant growth depends on six factors:

1. Texture
2. Structure
3. Consistence
4. Aeration
5. Tilth and
6. Moisture

SOIL TEXTURE

It refers to the relative proportions of sand, silt and clay particles in a mass of soil. The basic textural classes of soil are sand, loamy sand, sandy loam, loam, silt loam, silt, sandy clay loam, clay loam, silty clay loam, sandy clay, silty clay and clay. Sand particles vary in diameter from 2.0 mm to 0.05 mm, silt particles average in diameter from 0.05 mm to 0.002 mm. Clay particles are vary from less than 0.002 mm.

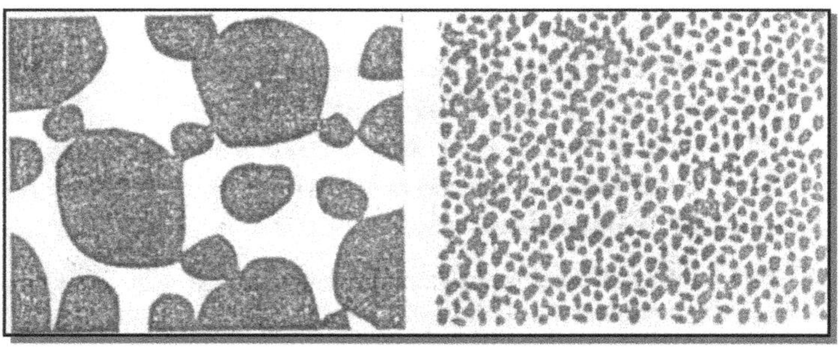

Fig. 9.1 : *Soil Texture*

Note: Individual particles of silt can be seen under the microscope, but individual particles of clay cannot be seen under the microscope. In nature, the soil is almost always a mixture of sand, silt and clay particles. Clay as a textural class may contain less than 40% silt, less than 20% sand and over 40% clay.

SOIL STRUCTURE

It means the arrangement of primary soil particles (sand, silt and clay) into compound particles or aggregate. The principal forms of soil structure are Platy (laminated), Prismatic (Vertical axis of aggregates longer than horizontal), Columnar (Prisms with rounded tips) Blocky (Angular or Sub angular) and Granular. Structure less soils is either single grained or massive (Table 9.1).

Table 2: *Soil Structure*

Classification	*Size*
Gravel	2 to 64 mm
Sand	0.05 to 2 mm
Silt	0.002 to 0.05 mm
Clay	Less than 0.002 mm

SOIL CONSISTENCY

Soil consistence is loose, friable, firm, plastic, hard and soft depending upon the characteristic of the soil texture and structure.

- **Loose:** when dry or moist, does not hold together in a mass.
- **Friable:** when moist, crushes easily under gentle pressure.
- **Firm:** when moist, crushes under moderate pressure.
- **Plastic:** when wet, readily deformed by moderate pressure.
- **Sticky:** when wet, adhere to other materials.
- **Hard:** when dry, moderately resistant to pressure, can be broken with difficulty.
- **Soft:** when dry, breaks into powder (under slight pressure).

SOIL AERATION

It relates to the exchange of air in soil with air from atmosphere. The air in a well aerated soil is similar to that in the atmosphere (21% of

Oxygen), while the air in a poorly aerated soil is several thousand times higher in Carbon-di-oxide (upto 20%) and lower in Oxygen because of excess water.

SOIL TILTH

It is the physical condition of the soil as related to tillage, seed bed preparation, seedling emergence and root penetration.

SOIL MOISTURE (SOIL WATER)

It is defined as the water held in soil pores between soil particles within soil granules. Soil water and soil air both occupy the same soil pores. The relative amount of each often fluctuates hour by hour. Available soil moisture is soil water held against the pull of gravity between the field capacity and the wilting point.

10
Chapter

Chemical and Colloidal Characters of Soil (Chemical Weathering)

Soils vary widely in their chemical and colloidal make up. This variation is due to the chemical composition and the parent materials and to the climate, plant and animal life, age of the soil. Nitrogen, Phosphorus, Potassium are important elements of the 16 essential nutrients for plants. Oxygen is the most abundant element in the earth's crust. Iron is the second most abundant in the soil in USA. Many soils that have developed from limestone are still have lime in the surface soil. Soils that have developed in drier areas are rich in lime because there have been very little loss of lime by leaching. Colloidal particles are clays and organic matter (humus) with particles less than 0.002 mm in diameter. Such particles are fine enough to carry a negative charge and to stay in suspension to make muddy water.

Acid and Alkaline Soils

The varying levels of minerals and acidity in the soil have a considerable effect different vegetation on it. An acid sandy soil, low in nutrients is favoured by coniferous trees and by plants that cannot tolerate much Calcium. These are known as calcifuges. If calcifuges are grown in alkaline soils, they suffer from poor iron metabolism. In contrast, calcicoles (calcium suckers) grow in Calcium rich alkaline soils often found in grasslands. Calcium and other alkaline compounds may accumulate in the soil in arid climates. They may be washed away by irrigation and prevent water logging.

CONSOIL, 2003: International Conference on Contaminated Land-Gent. Belgium, 12-16 May, 2003. (Email ID: consoil@fu-confirm.de).

11
Chapter

Biological Characters of Soil (Sustainable Agriculture)

The soil supports a large population of plants and animals. Among the important organisms and single celled bacteria and fungi, which vary from simple yeast cells to complex mushrooms, actinomycetes, thread bacteria and algae. Certain fungi and actinomycetes are the sources of antibiotics. In addition, there are many small animals such as protozoans, nematodes, earthworms, insects and arachnids in soils. Rodents are also present in soil burrows.

The majority of these microoganisms obtain their food and energy by breaking down complex organic substances manufactured by higher plants. Without microorganisms, human life cannot survive, since dead plants and animals would soon accumulated and form the earth's surface coating. Also, the nutrients in dead plants and animals are needed to recycle and feed vegetables directly and animals indirectly.

The various groups of organisms grow under quite different environment. Most bacteria prefer well drained, slightly acidic to alkaline soils with moderate temperatures and good aeration. Fungi prefer acid soils and Algae are found near the surface of the soil where light and water are available. While most groups depend on organic matter for food and energy, several types obtain their energy by oxidizing inorganic elements such as Sulphur and Nitrogen. They get their Carbon from the Carbondioxide of the soil atmosphere. A few organisms can live in poorly drained soils and anaerobic conditions.

The "Nitrogen-fixers" are important to farmers. Some groups are growing free in the soil and some are attached to the roots of certain plants called legumes. Both groups are able to take inert Nitrogen gas from the air and convert it into a form which plants can utilise. Some non legumes can also fix atmospheric Nitrogen.

The decomposition of organic matter in the soil by microorganisms completes a vital cycle in the ecosystem. The plant grows and absorbs nutrient elements from the soils, matures and dies. By microbial breakdown of the dead plant, these plant food elements are released again become available for the use of a new generation of plants. Without this decay process, most of the Carbon in our universe would become lockedup in a complex plant and animal tissues, since the amount of available Carbondioxide in the atmosphere is quite limited. When organice matter decays, various acids are formed in the soil. These may react with mineral compounds to make vital plant food elements such as Phosphorus, Calcium, Magnesium and Potassium more soluble and available to growing plants.

During the process of organic matter decay, an organic colloidal material that helps to cement the finer soil particles into larger water soluble granules is produced. A soil, which is granulated like this, is said to have good strucutre (tilth) and is often termed as "mellow" or easy to work. When it rains, the water soaks into a well granulated soil instead of running off the surface.

It should be emphasized that organic matter decay by soil microbes is a continuous process and fresh organic residues must be continually added to the soil if the maximum benefits of this complex process are to be beneficial (symbiotic) relationship with many plants, especially trees and shrubs.

Not all the activities of soil organisms are beneficial to plants and man, some groups produce diseases in crops. Certain organisms such as Nematodes attack the roots of plants, causing serious damage. Microorganisms may also complete with crops and plants for nutrients, especially for Nitrogen.

12
Chapter

Soil Organisms: Plants, Microorganisms and Animals

Without soil organisms, the earth would be covered with sterile mineral particles living soil ecosystems on which we depend for most of our food. The activity of the myriad organisms living in the soil help create structure, fertility and tilth (condition suitable for cultivation).

Usually soil organisms stay close to the surface. One hectare of their living layer contain thousands of species and innumerable of individual organisms. Algae live on the surface. Bacteria and Fungi flourish in the top few centimeters of soil. One gram of soil can contain hundreds of millions of these microscopic cells. Algae and Blue green bacteria capture Sunlight and make new organic compounds. Bacteria and Fungi decompose organic debris and recycle nutrients (Nitrogen cycle) that plants can use for additional growth. The sweet aroma of freshly turned soil is caused by Actinomycetes, Bacteria that grow in Fungus like strands, give us the Antibiotics such as Streptomycin and Tetracycline.

Among animals, the earthworms, roundworms (Nematodes), mites, tiny insects, rotifers, Barkudia, a limbless lizard, spiders, scorpions, crabs, centipedes, millipedes, snails, soil protozoans, sow bugs are present in the soil (Fig. 12.1). The earthworm is a farmer friend, which helps vermiculture for preparation of ecofriendly and effective manure. The largest soil dwellers are mammals, moles, shrews, mice and dogs are some mammals. Moles spend their lives within the soil, they never come out. Mice also never come out. They form a tunnel in the soil. Shrews spend almost as do moles, but they do see the light, come out. Mice spend much of their time above the ground. Air is present in the vegetation parts of soil and tunnels and burrows. Otters are found in sand barrows.

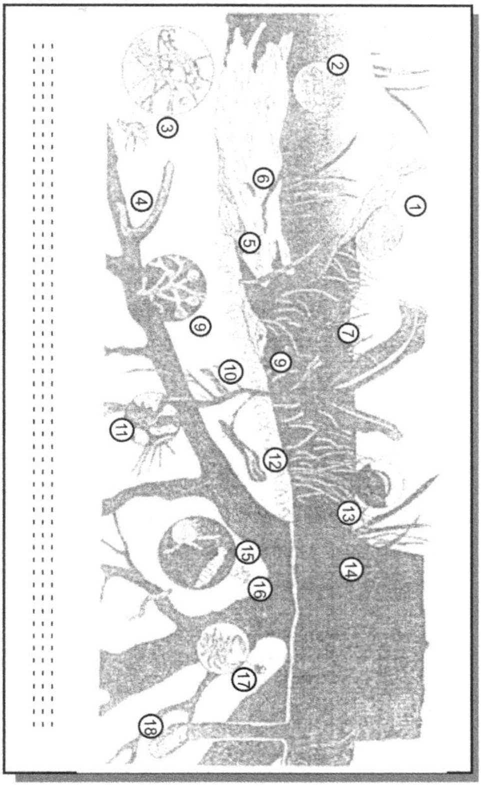

Fig. 12.1 : *Soil Organisms.*

Although mammals play a part in the making of soil, the most important soil animals are insects and others have very small creatures. There are more than a million kinds of insects. Most of them spend atleast part of their lives in the soils. Some of them as in eggs and others are in larval or pupal forms. Burrowing insects are live all their lives within the soil.

Other soil animals are earthworms, nematodes, sowbugs, centipedes, millipedes, spiders, scorpions and mites. Earthworms feed on soil and make soil much richer for growing plants. Earthworm is a farmer ecofriend. Scientists have estimated that there may be as many as million earthworms in an acre of soil. The earthworms pass as much 36 tons of soil through their bodies in one year.

Some of these soil animals (Fig. 9) are herbivores, but many of them pray upon one another. Soil roundworms (Nematodes) attack plant rootlets and can cause serious damage. Burrowing animals, such as moles, screws, insect larvae and worms which makes tunnel deeper in the soil, mixing and aerating it. Plant roots also penetrate into lower soil levels secreting acids that decompose mineral particles. Hence the same soil organisms help in maintenance of soil fertility through Nitrogen fixation and are responsible for return of essential nutrients. Fallen plant litter added new organic material to the soil. Thus nutrients are recycled.

LAND

It is estimated that there are 30 million species on earth, of which man is only one. Vertebrates are 41,000 and Insects are 750,000 species. Plants are 250,000 species which are in existence on earth now. Within the next generation, 1¼ of these species are at risk of extinction. Two major causes are the destruction of natural habitat including soil erosion and the hunting of animals for their commercial value.

FARM LAND

Protecting the valuable topsoil is necessary for outstanding food production. Crop rotation, Irrigation, Improved seeds and efficient farming methods and equipment are all vital for maintaining Farmland.

Land degradation occurs as a result of overgrazing, deforestation, agriculture and exploitation. Valuable topsoil is lost. In dry areas of the world, land degradation causes desertification.

Mismanagement of forests creates barren lands through deforestation and Podu (Shifting) cultivation.

13
Chapter

Soil Conservation and Management

Our earth is very precious as it creates favourable conditions for this existence of life. But due to the development of new technologies in various fields, through it improved the standard of living, the crucial environmental balance established by nature even millions of years have been inadvertently disturbed the need stares among the rations over the globe for development jeopardised the health of man itself. This craze resulted into unlimited exploitation of every bit of natural resources. Earth or land which is one of the natural resources is being deteriorated by anthropogenic means and it is attributed to three major causative factors. They are over population, urbanisation and industrialisation. The splendid plentifullness of nature is a heritage that should be conserved for future generations and not be spoiled.

Land is an important component of life supporting system. It consists of both living and nonliving as resources. Living resources constitute flora and fauna and nonliving resources constitute mineral deposits, water resources etc. These components are interdependent and interrelated upon each other and thus life is sustaining on this earth. Living animals on earth is possible only when flora (plants) are in balanced state as animals depend on plants for food, shelter and clothing. To maintain flora, fertility of soil is to be maintained. But due to the developmental activities and population outgrowth, more and more land is being taken for developmental activities such as construction and cultivational purpose, like constructions of roads, buildings, industries, factories and under cultivation programmes like agriculture and horticulture. Finally, the result of which is deforestation. According to the National Remote Sensing Agency (NRSA), India has been left with 22% forest cover. The land that have been taken for production purposes is loosing its fertility characteristics due to over exploitation of land and also due to addition of harmful pollutants into its which are released by various industries, sewage waters released from residential areas and leachates from agricultural lands. These pollutants change the physicochemical and biological processes that occur in soil (which are due to presence of micro fauna and flora) and may also make land a waste land or barren land and thus make the soil infertile for future use. To protect the land from becoming waste land are used to conserve the fertility of land, the main factors that degrade the land are:

1. **Deforestation:** India is losing 1.3 million hectares of land per year and the land is undertaken for constructional and cultivational activities to meet the demands of people.
2. **Soil Erosion:** Loss of vegetational cover due to over grazing of animals, the land becomes more susceptible to erosion. Agents of erosion like wind and rain water have left vast tracts of land barren. The splashes of rain water drops directly hitting the ground and then drawing or driving away the top soil which is rich in minerals to the near by rivers of streams, thus making the land infertile or waste land.
3. **Over Irrigation, Floods and Drought:** Over irrigation of land for cropping made the land saline and thus make land unfit for further cultivation. Floods and drought also lead the fertile land into waste land as microorganisms do not survive the water logged and severe dry conditions of land.
4. **Usage of Pesticides and Fertilisers:** Excess usage of pesticides and fertilisers into the soil leads to degradation of land.

Conservation of the soil from the above factors can be carried out by:

1. **Afforestation:** Declaration of forest land as Sanctuaries and National parks, waste lands or barren lands which can be used for afforestation. The plants selected for afforestation should grow fast and needs poor maintenance. Social forestry programme is needed to meet the demands of local people for fuel, fodder and timber, so that no more any large trees are cut. Discourage podu (shifting) cultivation.

2. **Soil Erosion:** It can be reduced by the following methods:
   (*a*) *Contour farming:* An oldest method useful in areas with low rainfall in the preparation of the field with alternate furrows and ridges. The water is caught and held in furrows and stored, which reduces run off and erosion.
   (*b*) *Mulching:* This method is effective against wind as well as water erosion. Some plants like maize stalks, cotton stalks, tobacco stalks, potato tops etc. are used as mulch (a protective layer formed by the stubble in the basal parts of herbaceous plants). Mulches (2-3" thick) reduce soil moisture evaporation and increase amount of soil moisture by addition of organic matter to soil.
   (*c*) *Crop rotation:* It decreases looseness of soil and preserves the productivity of land. The same crop year after year depletes the soil minerals. This can be over come by activating leguminous plants.
   (*d*) *Strip cropping:* It involves the planting of crop in rows of strips to check flow of water.
   (*e*) *Dry farming:* Growing grazing fields, animal husbandry are the possibilities of checking erosion in lands with moderate rainfall where farming is risk.
3. **Usage of Organic Manure Instead of Fertilisers:** Increase the fertility of the soil by providing moisture and nutrients necessary for cultivation as organic manure is formed of dead plants materials and animals materials which are rich in all minerals that are required for crop production.
4. **Treatment of Industrial Effluents before they Release on Land Surface:** Similarly sewage waters released from household areas must be treated by constructing sewage treatment plants, where the pollutants are reduced to a manageable level.
5. **Solid Waste Generated from Residents:** Industrial and Hospitals must be segregated according to their toxic levels and treated separately to reduce their toxic nature or by increasing them. So that harmful microorganisms do not spread into the atmosphere and soil non toxic material can be composed to get manure from it.

6. **Bioremediation:** It is a process which with the help of bacteria, toxic compounds can be converted into either useful industrial products or harmless wastes. This technique is useful in reducing the soil waste into fertilisers or manure.
7. **Biotechnology:** Biotechnological approach is recent technique with which the complex organic compounds which are non biodegradable can be decomposed by using genetically engineered bacterial strains.
8. **Recycling Method:** Papers, plastics and petroleum products which are main reason behind increase in wastes can be recycled. Papers on recyclation products has low grade paper, cardboards and packing materials which are useful again.

SOURCES OF SOIL POLLUTION

Soil pollution mainly the results of from the following sources:

1. Industrial Wastes
2. Urban Wastes
3. Radioactive Pollutants
4. Agricultural Practices
5. Chemical and Metallic Pollutants
6. Biological Agents

1. **Soil Pollution by Industrial Wastes:** Disposal of industrial waste is the major problem which is responsible for soil pollution. These industrial pollutants are mainly discharged from pulp and paper mills, chemical industries, oil refineries, sugar factories, tanneries, textiles, steel distilleries, fertilizers, pesticide industries, coal and mineral mining industries, metal processing industries, drugs, glass, cement, petroleum and engineering industries etc.

   With the advent of technology, new types of industrial wastes are produced and deposited in the land. These waste products are also tipped on soil, enhancing the extent of soil pollution. Thermal, atomic and electric power plants are also the **villainy** to add pollutants to the soil. The furnaces of such industries generate fly ash, i.e. unburn brownish black substance which severely pollute air, water and soil. Many industrial effluents are either discharged into streams or dumped into the surrounding land. Industrial wastes mainly consists of organic compounds along with inorganic complexes and non biodergradable materials. These pollutants affect the chemical and biological, properties of soil. As a result, hazardous chemicals can enter into human food chain from the soil or water, disturb the biochemical process and finally lead to serious effects on living organisms.

2. **Soil Pollution by Urban Wastes:** Urban wastes compress both commercial and domestic wastes consisting of dried sludge of sewage; all the urban solid wastes are commonly referred to as 'refuse'. Solid wastes and refuse, particularly in urban areas contribute to soil pollution. This refuse contains garbage and rubbish materials like plastics, glasses, metallic cans, fibres, paper, rubbles, street sweepings, fuel residues, leaves, containers, abandoned vehicles and other discarded manufactured products.

   Urban domestic wastes though they disposed off separately from the industrial wastes, can still be dangerous. This is so because they cannot be easily degraded. Over population and increasing consumption have totally changed the very complexion of domestic wastes into a complex mixture of food remains, paper, plastic and many notorious chemicals. Other items like paints and varnishes which we use to add colour to every day life also add poison to the urban wastes inturn poisoning soil pollution. The leachates from dumping sites and disposal tanks of sewage mixed with industrial effluents and wastes are extremely harmful and toxic. Actually the leachates that oozes out of the polluted soil, contain poisonous gases along with the partly decomposed organic material especially food remanents, vegetables, toxic hydrocarbons and pathogenic microbes many of which can be disease causing and harmful to man.

   Pollution concentration in urban areas and unplanned industrial progress in and around these urban areas, have a great extent contributed to soil pollution problems in India. About 12 crore population of India lives in cities while its six times more population lives in villages which dump their waste products into the soil, posing terrestrial and aquatic pollution hazard. It is estimated that Rs. 20 crore have been spent simply for burning, composition and thermally decomposing the refuse at Mumbai and Kolkatta only.

3. **Radioactive Pollutants:** Radioactive substances resulting from explosions of nuclear devices, atmospheric fall out from nuclear dust and radioactive wastes (produced by nuclear testing laboratories and industries) penetrate the soil and accumulate there and creating land pollution. Radio nucleotides of radium, thorium, uranium, isotopes of potassium (K-40) and carbon (C-14) are very common in soil, rock, water and air. Explosion of hydrogen weapons and cosmic radiations induce neutron-proton reactions by which nitrogen (N-15) produces, C-14. This C-14 participates in the carbon metabolism of plants which is then introduced into animals and man.

4. **Agriculture Practices:** Modern agriculture practices pollute the soil to a large extent. Today with the advancing agro-technology, huge quantities of fertilizers, pesticides, herbicides, weedicides and soil conditioning agents are employed to increase the crop yield. Many agricultural lands have now excessive amounts of plants, sand and animals wastes which are poisoning soil pollution problems. Apart from these farm wastes, manure slurry, debris, soil erosion containing mostly in organic chemicals are reported to cause soil pollution. USA alone produces about 18 million tons of agricultural wastes in every year.

5. **Chemical and Metallic Pollutants:** A number of industries including textiles, pesticides, paints, dyes, soaps and synthetic, detergents, tanneries, drugs, batteries, cement, asbestos, rubber, petroleum, paper and pulp, sugar, steel, glass, electroplating and metal industries pour their hazardous effluents in soil and water creating disastrous effects on living organisms.

   Synthetic chemicals and fertilizers are a source of trace metals which are added to the soil either deliberately or as an impurity. For examples, As, Pb and Cd, the common trace metals in rock phosphate, also occur in super phosphate fertilizers.

6. **Biological Agents:** Soil gets large quantities of human, animals and birds excreta which constitute the major sources of land pollution by biological agents. Digested sewage sludge as well as leave application of manures to soils without periodic leaching could cause chronic salt hazard to plants within few years. In addition to these excreta, faulty sanitation, municipal garbage, waste water and wrong methods of agricultural practices also induce heavy soil pollution. Sludges do have faults as they contain enough live viruses and viable intestinal works. In developing western countries, intestinal parasites constitute the most serious soil pollution problems.

CONTROL FOR SOIL POLLUTION

With the rapid pace of industrialisation and increasing population density, numerous pollutants have poised a serious threat to living organisms on this planet, extensive solid wastes and the use of biocides etc., have put the interest of agriculture and aquaculture at cross purposes. Some measures have been suggested to meet the challenges of soil pollution.

1. **Proper dumping of unwanted materials:** Excess of waste products by men and animals cause chronic disposal problems. Crude tipping or Open dumping is the most widely used method in Indian cities. Sanitary land fill can also be used but the process is not free from demertis. Recently controlled tipping is used in metropolitan cities like Delhi and Mumbai etc., for solid waste disposal. The surface so obtained then can be used for housing or sports fields.
2. **Production of Natural Fertilizers:** Organic wastes contained in animal dung can be used for preparing compost manure and biogas rather than throwing them wastefully polluting the soil.
3. **Use of Biological Weapons:** Recently a Committee of U.S. National Academy of Science have suggested that natural soil microorganisms should be used to increase crop fertility rather than the use of pesticides or insecticides. The committee have also recommended the formation of new generation of biological agents that can destroy the target insects.

4. **Proper Hygienic Conditions:** People should be trained regarding the sanitary habits. Lavatories should be equipped with quick and effective disposal machinery.
5. **Recycling and Reuse of Wastes:** To minimize soil pollution, the wastes such as paper, plastics, glass, metal organics, petroleum products and industrial effluents etc., should be recycled and reused.
6. **Ban on Toxic Chemicals:** Ban should be imposed on chemicals which are otherwise fatal to plants and animals (such as a few pesticides e.g., DDT, BHC etc.,) Nuclear explosions and disposal of radio active wastes should be banned.
7. To vitiate the insecticide hazard, the U.S. Committee have suggested the compulsory reporting of its formation and production to the Government.
8. Recently Industrial Toxicological Research Centre (ITRI), Lucknow has established central facility for safety Evaluation of pesticides to prevent their adverse effects of soil and crops.
9. Analytical laboratories should be equipped with the latest sophisticated instruments for detecting the micro quantities of pesticides in soil.
10. **Plantation of Trees:** Extensive plantation is extremely essential to prevent soil erosion. Some of the plants, Casuarina to protect marine soil erosion. These remedial measures in terms of effective and integrated ecological management of soil system are necessary to restore normal conditions.
11. **Unusual uses of Clay Soils:** In some areas, some people and cattle are eating clay soils. In Africa, cattle eat certain kinds of clay. Furthermore, some tribal groups plaster their bodies with mud, a practice that has now spread in parts of USA and India. All the earthworms eat soils, so that they are good friends of farmers.

SOIL EROSION

Soil Erosion is an important natural process, resulting in the redistribution of the products of geologic wathering and both soil formation and soil loss. The world's landscapes have been sculptured by Soil erosion process. There are many kinds of soil erosion. Based on the rate of soil loss, there are two main kinds of soil erosion.

1. Normal or Natural Geologic soil erosion and
2. Accelerated soil erosion.

1. **Normal or Natural Geologic Soil Erosion:** It occurs by natural or normal conditions by itself without involvement of man. Natural or normal soil erosion, it is very slow process and equilibrium between loss of soil and builtup soil is hardly lost, unless there is some major disturbance by foreign agent.
2. **Accelerated Soil Erosion:** This type of removal of soil is rapid and never keeps pace with the soil formation. It is a most serious type of loss, generally caused by an interference of an agency like other animals and man.

Agents of Soil Erosion

Any natural agency that is able to transport the weathered rock material is known as an agency of erosion. The common soil agents are ground water, running water, moving ice crystals, wind and sea waves. Glaciers are important agents in polar regions wind action prefers in desert areas. Soil erosion differs from one season to other and place to place depending upon the slope force of gravity in moving soil particles. An agent of soil erosion carves out the landscape characters like valleys, ridges and hills and continuously modifying them.

Machanism of Soil Erosion

Based on the agents of soil erosion, different kinds have been recognized. They are Sheet erosion, Rill erosion and Gully erosion. In Sheet erosion, the soil is removed by water in the form of their covering from over large area. The sheet is lost more or less uniformly. In Rill erosion, soil is removed with full force, the run off water moving rapidly over the soil surface cutting finger like grooves, appearing as their channels or streams called as rills. In Gully erosion, became still wider and deeper due to heavy rains. Moreover, several rills coverage towards steep slope forming wider channels of water gullies.

Slip Soil Erosion of Land Slide

Due to hydraulic pressure caused by heavy rains increases the weight of the rocks at cliffs that passes under the gravitational force and finally slip down. This is called as land slide or slip soil erosion.

Riparian Soil Erosion or Stream Bank Erosion

During floods, the rivers splash their water against the river banks cutting through them particularly at curves, wate strikes with great speed and the bank of river caves in alongside.

In case of wind erosion, salty soil carried in the form of small leaps is known as saltation. The small particles of soil blown up by wind as fine dust is called as suspension. It may be surface creep, when heavy particles of soil are simply pushed or spread along the surface by wind.

Control of Soil Erosion

There are three methods to control of soil erosion. They are:

1. Biological methods
2. Mechanical methods
3. Other methods

1. Biological Methods: By the use of plant or vegetation cover.

A. *Agronomic Practices:* Natural protection by growing vegetation in a manner that reduces soil loss.

(*a*) *Contour farming:* Oldest methods useful in areas with low rainfall. It is the preparation of the field with alternate burrows and ridges known as contours.

(*b*) Mulching: It is effective against wind as well as water erosion. Some such plants as maize stalks, cotton stalks, potato tops etc., are use as mulch.

(*c*) *Crop rotation:* It decreases soil loss and preserves the productivity of land. The same crop year after year depleted the soil minerals.

(*d*) *Strip cropping:* It involves the planning of crop in rows or strips to check flow of water. It may be contour strip cropping, so that the water used is minimal.

B. ***Dry Farming:*** The practice is useful for crop plants grown in low and moderate rainfall areas where ordinary farming is at risk. Crop production, animal husbandry and growing grazing fields are the only possibility of checking the soil erosion.

C. ***Agrostological Methods:*** Grasses such as *cyonodon dactylon* are utilized as erosion restricting plants. They act as stabilizers when grown in fully and sodding.

   (*a*) *Lay farming:* This aims to grow grasses in rotation with the field crops which helps in building up the structure of soil, preventing soil erosion and improving its fertility.

2. **Mechanical Methods:** These methods are used as supplements to biological methods. They are:

   A. ***Basin Lifting:*** To construct small basin along the contours to retain water, which also reduces its velocity.

   B. **Contour Terraining:** To construct a channel along the slope to intercept and divert the runoff water. This may be:

      (*a*) *Channel Terrace:* To dig channels at suitable intervals and the excavated soil deposited as a wide, low ridge along the lower ridge of the channel.

      (*b*) *Broad based ridge Terrace:* To construct ridge along both the sides of the channel.

      (*c*) *Bench Terrace:* To construct a number of platforms along contours or suitable graded lines across the slope.

3. **Othe Methods:**

   A. **Gully Control:** By constructing bunds, dams, drains or diversion channels.

   B. **Protection from Stream Bank:** The growth of the vegetation along the side of the riverbank. The construction of drains, concrete or store embedding etc. for checking the cutting and caving of riverbanks.

   C. **Afforestation:** Trees like Casuarina, mangroves, badam, soap nut trees and cashew trees are planted along shores which control the soil erosion and reduce the Tsunami and other heavy floods.

14
Chapter

DESERTIFICATION OF SOILS

India is extending over 329 million hectares of land. Out of this land, 24 mha of land is inaccessible under Foreign occupation; 18 mha of land is urban and other nonagricultural use; 21 mha is barren land. Only 266 mha of land is available for potential use (Table 14.1).

Table 14.1: *Land used for different practices in India*
Total land area: 329 million hectares

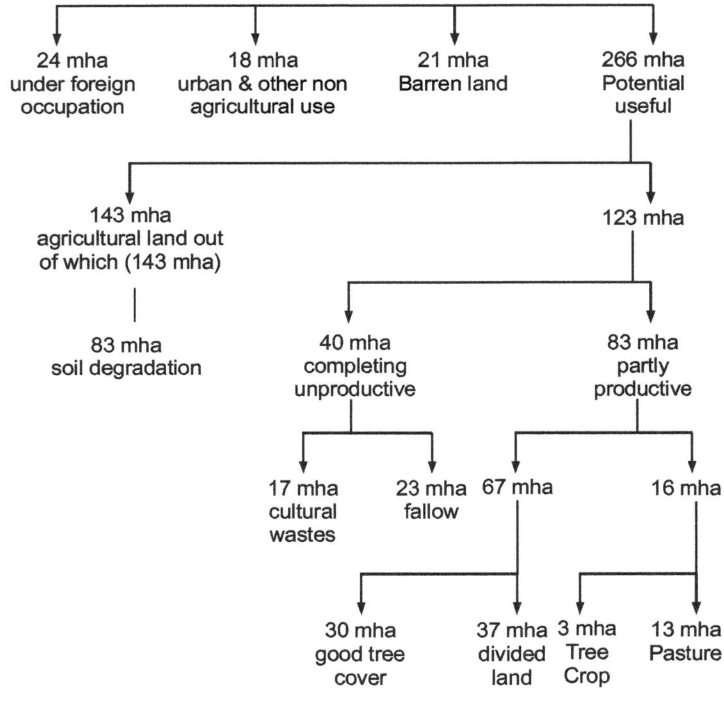

This potential land (266 mha) in India is divisible into 143 mha (more than 50% is an agricultural land. This agricultural land have suffered 83 mha for soil degradation. 40 mha land is completely unproductive, and remaining 43 mha land is partly productive.

According to India's National Forest Policy (1952), 33⅓ % of land should be under forest cover for ecological balance and sustainable development but at present, the forest cover in India is about 21%. India is losing 1.5 mha of forests per year; thereby bringing down the total forest areas from 74 mha to 40 mha (NFP Report, 1999).

Deforestation causes loss of top soil to the tune of 12,000 million tonnes per year. Due to this, India loses Rs. 10,000/- crores per year, in the form of floods and cyclones. Large scale felling of trees causes irreparable loss of organic sources. India requires 80 million cubic meters of fuel wood in every year. The Government of India has promulgated the Forest (Conservation) Act in 1980 to control indiscriminate deforestation/diversion of forest land to non forest purposes. During Independence of India (August 15, 1947), the population was 33 crores. In 1982, the population in India was 65 crores. Now (2009), the population in India in more than 118 crores. Within the 62 years, the population increases more than three times (Table 14.2). Now heavy pressure on population in India which results the crisis of food, house and commodities etc. Only solution is to check the population as well as the deforestation to save our mother land in an amicable way. Hence protect our environment and conserve our natural resources including the soil.

Table 4: *Population in India and population pressure*

Independence day 15th August	*Population in India*
1947	33 crores
1982	68 crores
1992	85 crores
2000	100 crores
2005	106 crores
December 2008	116 crores
December 2012	120 crores

- World's population in 1992 is 5,293 millions.
- China is the first largest population country in the world (1,160 million population).

- India is the second largest population in the world (840 million population).
- Every year in India, 1.3 crores of population increases.
- India's land area is 3.29 million Sq. km.
- Worlds land area is 135.79 million Sq. km.
- Population growth in India (1981 to 1991); 23.50%; Density; 267/1 Sq. Km. and; (1991 to 2001) 28.75%; 355/1 Sq. Km. respectively.
- Environmental problems in population are:
  1. Land degradation
  2. Desertification
  3. Loss of Top soil due to soil erosion
  4. Depletion of natural resources
  5. Almost disappearance of rare species of plants and animals
- Population pressure effects 22 reasons:
  1. Literacy
  2. Hunger
  3. Water
  4. Housing
  5. Unemployment
  6. Income
  7. Urbanisation
  8. Overgrazing
  9. Climate changes
  10. Natural Recreation
  11. Marine Fisheries
  12. Pollution
  13. Inflation
  14. Environmental disorders
  15. Over crowding
  16. Deforestation
  17. Political conflict
  18. Minerals
  19. Health
  20. Energy
  21. Endangered species
  22. Individual freedom

15

Chapter

GENERAL SOIL MAP

Soil maps are accompanied by soil profile description with soil survey area of published soil survey report with properties of soils and significant uses are available at district agriculture office in each state.

The area of the earth that is suitable for growing food is limited. Ice, snow, desert and mountains covers 51% of the land. Most of the world's population is fed by another 21% of the land. As population grows and food production expands, this land is under ever increasing pressure, as much as 30% of agriculture land around the world is at risk from over use. In developing countries, 65% of today's agricultural land may become unfit for farming in the coming two decades.

Agriculture consumes huge quantities of water (about 41% of total use in USA and 87% in China). About 40% of people in the world suffers frequent severe droughts. Competition for access to water is increasing within states in India and with other countries. An increasing number of dams and reservoirs are being built in an attempt to regulate water supplies. However, reservoirs are being filled with silt overtime and the flooding of land disrupt ecosystems and displace large number of people.

LANDS IN INDIA

The total land area of India is 32, 87, 263 Sq. Km or 328.7 million hectares extending from snow covered Himalayan heights to tropical reinforests of the south. Out to the total land area, 43% is covered by plains, 28% is plateau and 29% comprises of mountains and hills. In all, only 62% of the total land area is usable. 24 Mha is inaccessible under foreign invasion, 18 Mha of land is available for potential use (Table 3). This potential land, 266 Mha in India is divisible into 143 Mha (more than 50%) is agricultural land. Out of this, 83 Mha suffers from soil degradation, 40 Mha land is completely unproductive and remaining 42 Mha land is partly productive (Table 3). India is losing 1.5 Mha of forests per year, thereby bringing down the total forest area from 74 Mha to 40 Mha. Deforestation causes loss of top soil to the tune of 12,000 millions tons per year.

The three major components of our lands are crop, forest and grazing lands. Out of these, grazing lands have been most neglected. The environmental problems created by mostly neglecting the grazing lands in India clearly shows that the current compartmentalized management of India's forest crop lands by departments of Forests and grazing lands by the department of Animal Husbandry, none of which talk to each other or care to understand each other's problems. This can only lead to further deterioration of lands in India.

They are also being destroyed by indiscriminate mining activities with little care being taken to restore these lands. The excessive pressure on grazing lands is the absence of organised fodder production and pasture management. The growing human population has increased pressure on lands to produce more food grains and raise cash crops. Forage crops do not find an important place in crop rotation because of the small holdings.

LANDS IN THE WORLD

Of the three, main components of the earth system are Atmosphere; Water and Land. The land is most heterogenous. The earth's surface is a mosaic of different types of ecosystems ranging from arid desert to tropical forests to Tundra to more familiar temperate forests. Each harbors distinct plant and animal communities. Each contributes unequally to the functioning of earthsystems. The total land area of the World is 13,128,841 Hectares. The crop land is 1,47,877 Ha; permanent pasture is 3,322,943 Ha. The forest and wood land in the World is 4,232,737 Ha and wild area is 3,486,997 Ha. The population density in the world is 403 per 1000 Ha as per 2001 census (Fig. 10)

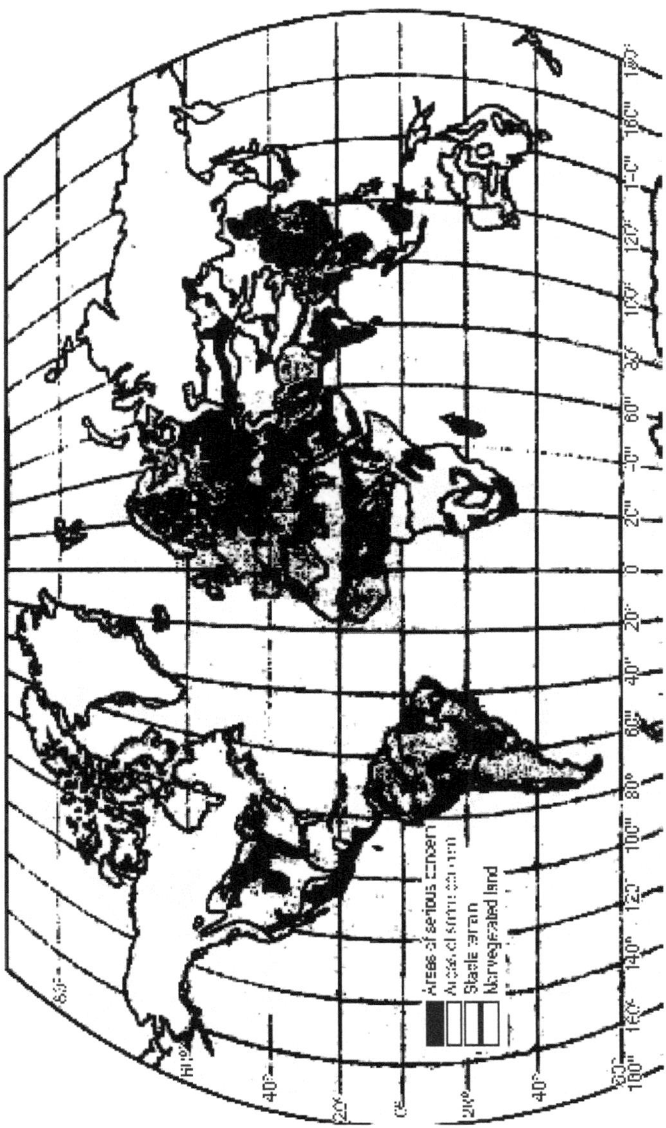

Fig. 15.1 : *Areas of concern for soil degradation. In areas of serious concern, widespread moderate or localized severe degradation has already occurred, and resources are lacking Environmental conditions make rehabilitation difficult.*

UNIT-II

16
Chapter

SOIL ANALYSIS

INTRODUCTION

Soil

Soil is a thin layer of material that lies over rocks, covering most of the land. It may be only few centrimeters deep, or it may extend several meters down to the rock below. Soil forms the link between the abiotic and biotic parts of a terrestrial ecosystem. Plant roots grow through soil and take in water, minerals and Oxygen. It has four main components. Mineral particles (upto 60%), organic matter (about 10%), water (upto 35 %) and air (upto 25%). The mineral particles of soil are derived from underlying rock as it undergoes weathering.

Physical weathering can be caused by changes in temperature that cause the rock to expand and contract, weakening it so that it eventually shatters. Plants such as mosses and lichens may grow through the crocks, loosening the rock material. Further breakdown occurs in chemical weathering as the rock is exposed to Oxygen from the atmosphere or acids in rainwater. Bacteria, Fungi and Lichens are also contribute acids from chemical weathering.

The mineral particles in soil are distinguished by the size. Sand is the largest than silt and the smallest one is day. The proportions of each of these components give a soil and its particular characteristics. A soil with a lot of sand and little clay is light weight with many air spaces and drains easily, but it is poor in nutrients. A soil with mostly clay particles in much heavier, holds more water and is slower to drain. A loamy soil with balanced amounts of sand, silt and clay, is best suited for agricultural use.

The character of a soil also depends upon its chemical composition, which derives from the rock from which it formed. A sandy soil may also carry layers of iron or Aluminium oxide (known as Podozol), salty soils with a high proportion of Sodium and clay rich subsoil (solonetz) are frequently found in arid regions.

The organic matter that accumulates as the top layers from humus, dead material such as fallen leaves and the remain of animals. Humus gives soil its dark colour and nutrients, and improves water retention (in sandy soils) and drainage (in clay soils). Bacteria in the humus play an important role in fixing atmospheric Nitrogen and making it available to plants. New soil contains no humus. Mature soil takes as long as 10,000 years to develop, while plant cover grows to allow nutrients to circulate between the soil and vegetation. If it is not over exploited, the soil remains fertile for millions of years. Without plant cover, the soil becomes badly eroded within decades and cannot be replenished.

ORIGIN AND NATURE OF SOILS

The United States Department of Agriculture (1951) defined soil as "the collection of natural bodies occupying portions of the earth's surface that support plants and that have properties due to the integrated effect of climate and living matter acting upon parent material, as conditioned by relief, over periods of time."

Of many different kinds of soil on earth, each has its own special collective characteristics upon which its behaviour is depended. It is necessary to distinguish between soil as an important substance and soil as a natural body.

The earth's crust is the origin of soil. On the process of weathering, primary rocks produce what is known as "Soil parent material" and the rocks from which this material is produced called "Parent rocks". Soil parent material are now recognised in four main classes:

I. Those formed in situ from soft rocks such as chalk and volcanically.

II. Those formed insitu from hard rocks which may be igneous (e.g., granite and basalt), sedimentary (e.g., limestone and sandstone) or metamorphic (e.g., genesis and marble).

III. Those transported from their place of origin before weathering to form soils. This is the most important of the parent material classes and includes all material moved by water (such as alluvium, colluvium, lacrustine deposits and beach sediments), by wind (such as loose and ecolian sands), and by glacial phenomena (such as glacial drift and tilt).

IV. Those formed from organic material such as peat.

Rocks are weathered mechanically by the action of frost, water, wind and chemically by such processes as oxidation, hydration, hydrolysis and solution in Carbonic acid. Similar or agential rocks may produce quite different materials basing on the kind of weathering.

These products undergo further change to produce soil, as we know it. These processes do not necessarily occur at the same time or place as production of parent material as the soil forming processes may have proceeded concomitantly with rock weathering. In many soils as now found, more than one parent material may have been involved in their formation.

For a better understanding, the process of soil formation from parent material can be divided into two phases. The first phase involves addition and admixture of organic matter and the other involves solution and translocation of certain constituents, which is known as 'eluviation'. Eluviation itself can be further divided into mechanical eluviation in which finer particles, usually move downwards by washing and chemical eluviation which involves the partial decomposition of colloidal matter with subsequent downward movement of the decomposition products.

During or after the soil formation, its nature can be modified by unrelated effects such as volcanic action, floods, erosion and human activities.

The eventual soil is chemically, physically and biologically complex, dynamic system, the constituents of which are constantly undergoing change. Hall (1949) likens soil to a three phase system of solid, liquid and gaseous components with constantly shifting equilibrium.

For further understanding and discussion on the origin of soils, it is recommended to refer to the standard works on soil genesis and constitution (Hall, 1949; Russell, 1961). From the analyst's point of view, it is apparent that soil, even as material in a bottle, provides a stimulating challenge to his skill and capabilities.

SIGNIFICANCE OF SOIL

Soil is one of the most significant ecological factors, which is derived from the transformation of surface rocks. It is nothing but soil on which plants depend for their nutrients, water and mineral supply and anchorage. It constitutes an important medium wherein numerous animals live in. In fact, soil of any nation is its most valuable material heritage.

The living material in soil is called humus, water, air and bit of rock are the nonliving materials that makeup soil.

All soils are mixture of tiny pieces of rock, water, air and humus. Different amount of these soil ingredients make different types of soil. Based on the amount of sand, silt or clay, it contains the type of soil. These soil ingredients help to give soil colour, usually darker (more humus) than lighter coloured soil (less humus).

SOIL NUTRIENTS

Soil contains nutrients called minerals. Plants need only small amount of these nutrients. If they get too much minerals from soil, plants can die. If they don't get enough minerals, plants cannot grow properly. Minerals come from the two soil ingredients i.e. tiny pieces of rock and the humus. Manure is a good source of Nitrogen. Some fertilisers come from nature. For e.g. plants and animals add nutrients from other fertilisers, made in factories. Earthworms helps the waste materials into the manure. Hence earthworms are ecofriends of the farmers.

Type of Nutrient

1. **Nitrogen:** Helps plants to grow. Helps plants to stay green.
2. **Phosphorus:** Helps to form flowers, fruits and seeds. Helps roots to grow.
3. **Potassium:** Helps plants to resist diseases, make seeds healthier.
4. **Calcium:** Helps plants to absorb other nutrients they requires. Helps roots and leaves to grow.
5. **Magnesium:** Helps plants use Sunlight to make food. Key part of preparation of chlorophyll.

PURPOSE OF SOIL ANALYSIS

Soil Analysis is one of the primary requirements for any land use management to a plant or crop, to construct a road, to drain, to fertilize and so on; or it may be analysed to discover its origin or to see if a certain constituent is present or to investigate the effects of a changing environment.

In the study of soil as a nutrition factor, for example the analysis may be of a routine nature for advisory purposes, or highly specific as in basic research. For the classification of soils as in a soil survey, again the analysis may be general in nature, but if a soil map is required illustrating a particular soil property such as salinity, then specific analysis of soil in its capacity as a medium for plant growth but one should realise that factors other than the chemistry of a soil such as, climate, microbiological conditions, structure and other physical aspects, may influence the plant growth. These factors are briefly touched upon here.

For whatever purpose a soil is to be analysed, plant should be made in advance as to what analysis are required, why they are required and what level of accuracy is needed. This is particularly important when an investigator has the analysis done for him rather than performing them himself, as more often than not a researcher demand an excessive number of analysis chosen from this book rather than from the requirements of the investigation.

SAMPLE COLLECTION

A variety of equipment is available for collection of soil samples depending upon the nature of the soil and the type of investigation. If weight is the basis, a trawl may be sufficient but samples whose volume must be known are normally taken with a graduated metal cylinder. Push the cylinder into the soil which helps in taking a core of soil of different depths. The compression of soil core can be minimized by using detachable cutting head and also enabling the core to be removed more easily. This is an advantage if the core has to be sectioned.

As stainless steel cylinder is the most durable an insert one, it can safely used in taking soil cores. A simple core which is effective for peat and clay soils can be made by rolling thin sheet metal (about 22 gauge) and fastening along the overlap to give the cylinder. After sampling, the fastening device can be released, enabling the cylinder to spring apart giving access to the core. For all corers, it is necessary to attach a horizontal 'tummy bar' to give a twist leverage during insertion into the soil.

Soils which are permanently water logged or where the water table oscillates considerably are always difficult to sample if some meaningful results are to be obtained. Adequate replication in the field is essential particularly as sub sampling at a later stage for various analysis may bring in additional errors.

GRINDING AND SIEVING

When a soil is air dry, it is taken to the preparation room for further treatment. The preparation room should be quite separate from the drying room, and particularly from the laboratory, and should have some means of dust extraction to prevent contamination of other samples.

It may be necessary to keep a part of the sample as a reference, or for setting up an exhibit or monolith, and such a sub-sample be preserved before any further treatment is given to the soil.

Stones, pebbles and pieces of macroorganic matter taken out and, if necessary, weighed, and the remaining sample is crushed by using a wood roller or a mechanical grinder.

After grinding, the soil is allowed to pass through a 2 mm (10 mesh). Approximately 1000g of fine earth is selected by quartering method (Fig. 16.1). The greater than 2 mm soil remained on the sieve is weighed and noted.

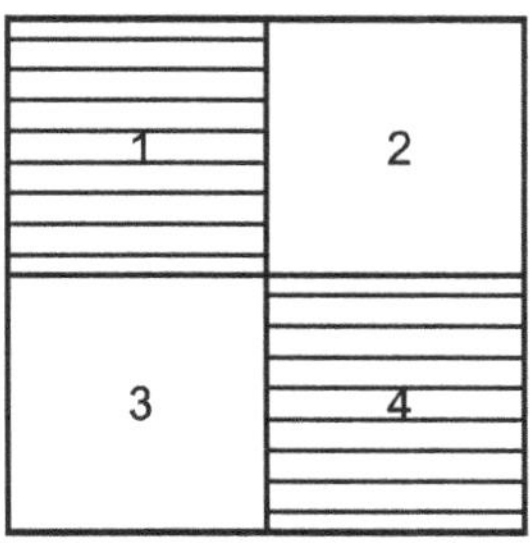

Fig. 16.1: *Reducing a soil sample by Quartering*

In 'quartering', the soil is spread uniformly over a sheet of brown paper or polyethylene and divided into four equal quarters as illustrated in Figure. 16.2. The quarters marked 1 and 4 are discarded and the remaining portions mixed together, spread out again and reduced to half by the same procedure. This process is repeated until a sample of the required bulk is obtained.

For certain analyses, in particular those carried out at micro or semimicro level, a sample much finer than the 2 mm is required. In some laboratories it may, however, be more desirable to prepare the sub-sample as and when required. In either case, the whole of the 2 mm sample is tipped out on a sheet of glazed paper or plastic sheet and sampled by quartering until the required amount is obtained (usually about 100g). The selected sub sample is then grounded in an agate mortar, either by hand or electrically, to pass a 0.5 mm (40 mesh) sieve or, for certain specific purposes, fine sieves. The complete sub sample must be ground to pass the sieve and none is discarded.

The composition of the grinding and sieving apparatus can be important especially if trace elements are to be determined. If minor elements are to be measured, extra precautions, against contamination of the sample must be taken throughout its preparation.

PHYSICAL CHARACTERS OF SOIL

A. Temperature

Soil temperature is measured by using soil thermometers. The bulb is buried under the soil to the desired depth and the temperature can be read on the dial.

B. Soil Colour

Soil colour is an index of soil fertility and has been the basis of naming big soil groups. It also helps in distinguishing soil profiles. Soil colour is mainly due to the presence of metals like iron, manganese and organic matter.

The Munsell soil colour chart contains standard colour chips arranged systematically on cards carried in a loose note book. Three simple variables, namely hue, value and chroma combine to to give almost all the colours. Hue of a colour indicates its relation to spectral colours like red, yellow, green etc., value refers to its relative lightness, and the chroma gives its strength (or departure from the neutral of the same lightness).

Spread the soil uniformly over a cardboard and select the card which contains matching colour chips. Each card carries colour chips of constant hue, designated by the symbol (letter observation of the spectural colour) in the upper right hand corner of the card. Each vertical row shows colours of increasing value. In horizontal rows of colours increases in chroma towards right. The nubers printed below each colour chip indicate its value and chroma respectively. Locate the nearest matching colour chip. In case of variegated or mottled soil the dominant colour chip determiner the change of colour. In writing the Munsell rotation, order is hue, value, chroma with a space between the hue letter and succeding value number, and a virgule between the two numbers for value and chroma. Record the colour name and Munsell rotation for dry and moist soils are as given below :

Reddish Brown (5 Yr 4/4; 3/4; moist)

Brown (10 Yr 5/3; 4/3; moist)

C. Soil Texture

Soil texture refers to the relative proportion of the various groups of individual particles. Particles more than 0.2 mm in diameter, viz. sand and gravel mainly provide physical support to the plants. Amount of silt and clay particles (less than 0.02 mm dia) responsible for water holding capacity and nutrient availability of the soil. Soils are classified into a number of textual classes depending upon their particle size distribution. Texture of the soil can be evaluated approximately in the field or accurately in the laboratory by mechanical analysis.

Soil sample is examined under hand lens and felt between figures in dry as well as moist state.

Table 16.1 : *International System of Textural groups in soils*

Textural group	*Size in mm*
Coarse	2.00 – 0.20
Fine sand	0.20 – 0.02
Silt	0.02 – 0.002
Clay	Below 0.0002

Examine the soil under hand lens for large particles. Squeeze small amount of air dry soil as well as moist soil between thumb and finger and try to judge the texture with the help of a key given below:

Sand: Sand is loose and coarse grained. The individual grains can readily be seen or felt. Squeezed when moist. It will form a cast, but will crumble when touched.

Sandy Loam: A sandy loam is a soil containing much sand but which has enough silt and clay to make it somewhat coherent. The individual sand grains can readily be seen and felt. Squeezed when dry, it will form a cast which will readily fall apart, but if squeezed when moist, a cast can be formed that will bear careful handling without breaking.

Loam: A loam is a soil having a relatively even mixture of different grades of sand, silt and clay. It is mellow with a some what gritty feel, yet fairly smooth and slightly plastic. Squeezed when dry, it will form a cast that bear careful handling, while cast formed by squeezing the moist soil can be handled quite freely without breaking.

Silt Loam: A silt loam is a soil having a moderate amount of the fine grades of sand and only a small amount of clay, over half of the particles being of the size called 'silt'. When dry it may appear cloudy but the lumps can be readily broken, and when pulverized it feels soft and floury. When wet, the soil readily runs together and puddles. Either dry or moist it will form casts that can be freely handled without breaking, but when moistened and squeezed between thumb and finger it will not 'ribbon' but will give a broken appearance.

Silt: A silt is 0.02 to 0.002 mm size of soil.

Clay Loam: A clay loam is a fine textured soil which usually breaks into clods or lumps that are hard when dry. When the moist soil is pinched between the thumb and finger, it will form a thin 'ribbon' which will break readily, barely sustaining its own weight. The moist soil is a plastic and will form a cast that will bear much handling. When needed in the hand it does not crumble red but tends to work into a heavy compact mass.

Clay: A clay is fine textured soil that forms very hard lumps or clods when dry and is quite plastic and usually sticky when wet. When moist, soil is pinched between the thumb and fingers it will form a long flexible 'ribbon'. Some fine clays very high in colloids are friable and back plasticity in all conditions of moisture (Fig. 16.2).

Space Front: The clay should feel slippery. The sand should feel rough and gritty. Clay and Sand are the major mineral solids found in soil. Clay and sand are made by the natural break down of minerals and rocks. Clay comes mainly from the mineral fieldspar. Sand comes mainly from the rock such as dry-rot which contains quartz. Both clay and sand contains silica. Silica is a chemical substance made of atoms of silica and oxygen. Sand is mostly silica because sand comes from quartz, which is nearly pure silica. Sand grains ranges from 0.062 to 2.0 millimeters. Clay is silica that contains Aluminium atoms and water. Other metal atoms are also found in clay. Grains of sand have many different shapes. Round shapes sand grains have smooth corner which carries water but sand grains have sharp edges which does not carry water. Clay is used to make ceramic objects such as bowls, plates and flower pots.

Geologists use the term silt which soil particles are larger than clay particles but smaller than sand. The diameter of silt particles are between 0.002 to 0.062 millimeters. Silt particles can be either mostly silica like sand, or can contain the same chemical substance found in clay. Clay feels support when it is wet (Fig. 16.2).

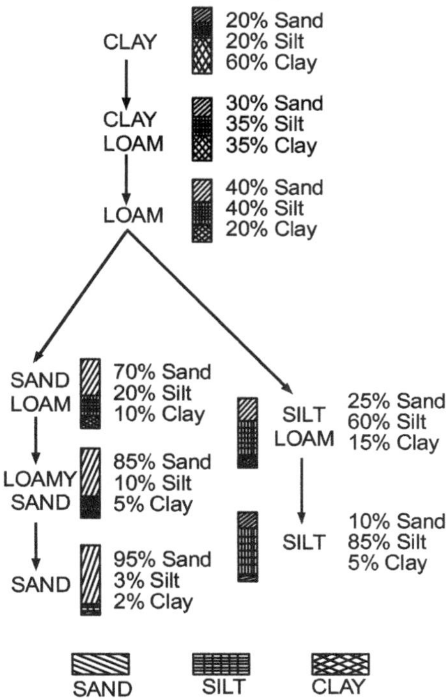

Fig. 16.2: *Soil Texture : Percentage of Clay, Clay Loam and Loam Soils*

Bulk Density

Bulk density, or apparent density, is the ratio of the mass of the soil to that of its total volume. The mass can readily be determined by drying to constant weight but the volume is less easy to measure accurately.

The volume of the soil can be known by using a metal cylinder. The walls of the cylinder should be as thin as practicable to minimize any compaction. Certain light, sandy and organic soils are difficult to extract in this way but can be removed after driving a plate horizontally to seal off the cylinder base.

Double wall cylinders are very effective for removing soil cores. In these, the inner cylinder containing the sample may be with drawn separately and the excess soil can be trimmed off cleanly at each end. Single cylinder should, if possible, be driven in slightly below the soil surface to allow for trimming the ends.

Cube blocks can be prepared in situ from some soil types, particularly amorphous beats and clays. If stones or roots make it impracticable to use the above method, it may still be possible to obtain an irregular block. This should first be weighed and then coated with paraffin wax of known weight and specific gravity. The volume of the block obtained by water displacement. Separate soil samples should be taken for moisture determination. A frozen soil block also be used to give an approximate result.

DETERMINATION OF BULK DENSITY (BD), PARTICLE DENSITY (PD) AND % PORE SPACE (PPS)

Aim

To determine bulk density, particle density and % pore space of a given sample.

Requirements

Soil block of definite size, equipment for weighing.

Procedure

Soil core or soil block should be taken of definite size and weight of the soil block is taken which gives total weight. The volume of soil sample should be noted. Then the sample is dried in, not air oven for 4-6 hrs. at 0°C. Then dry weight of sample is taken as H_2. By dispersing the soil is known volume of water and the volume of solids (Vs) is found out by using the above data, Bulk Density (BD) Particle Density (PD) and % Pore Space (% PS) is determined by using the following formula :

$$\text{Bulk Density (BD)} = \frac{Dry\ weight}{Total\ volume} \times gm / cm^3$$

$$\text{Particle Density (PD)} = \frac{Dry\ weight\ of\ soil\ sample}{Volume\ of\ solids} \times gm / cm^3$$

$$\text{Percentage Pore Space (PPS)} = 100 \left(1 - \frac{B-D}{P-D}\right)$$

Calculation

Calculate per cent pore space by the formula given below :

$$\text{Percentage pore space}: \frac{2.6 - D}{2.6} \times 100$$

Where D is the volume weight of soil and is calculated as oven dry weight of soil sample removed from the pit divided by the volume of the pelt, and 2.6 is the approximately specific gravity of soil particles.

Result

$$BD = 2\ gm/cm^3$$
$$PD = 2.5\ gm/cm^3$$
$$PPS = 20\%$$

Conclusion

The BD includes air space where as PD does not involve air spaces, PD-BD. A soil block of dry wt of 500 gm when allowed to future with water so the started weight of the soil come to 650 gms. Find the water holding capacity (WHC).

$$WHC = \text{Sat. Wt.} - \text{Dry Wt.}/\text{Dry Wt.} \quad 100$$
$$= 650 - 500/ 500 \quad 100 = 30\%$$

SOIL MOISTURE CONTENT

Aim

To determine the soil moisture content (MC) of a given amount of sample.

Requirements

Different soil samples, Balance, Weight box and Hot air oven.

Procedure

30 gm of soil sample is taken and made into powder with the help of a Pestle and mortar. This fine powder is placed in hot air oven for about 10 hrs. After heating, sample is weighed. This weight is substructure from the initial weight to get the moisture content of the given sample. This procedure is repeated for different soil samples. From the results obtained, the moisture content of the soils are calculated by using the following formula :

$$MC = \frac{FWT\text{-}DWT}{DWT} x100 \text{ where}$$

FWT = Fresh weight of the soil

DWT = Dry weight of the soil

Discussion

Soil is a complex components. It is madeup of different components, mineral matter, soil, air, soil water, soil solution and also soil organic matter, humus and micro organisms. Soil have low moisture content because of their large sized particles. Therefore evaporation is high. Silt soils have relatively small as well as penetration and percolations are high pore spaces than the sandy soils. Clay soils have highest moisture content because of this smallest pore spaces, which are more in number. The soils have low perculation and evaporation.

Calculation

Fresh Weight of the soil in soil cube (FWT) = 550 gm

Dry Weight of the soil in soil cube (DWT) = 500 gm

After 24 hours and cool it

$$\text{The moisture (water) content (MC)} = \frac{FWT - DWT}{DWT} x\ 100$$

Result

Total Moisture (water) content = 10%

Precautions

1. After fresh weight, keep it hot oven for over night.
2. Cool the content and weigh for dry weight.

3. Calculate as per the formula for percentage of the moisture content.

$$MC = \frac{\text{Fresh Soil Weight} - \text{Dry Soil Weight}}{\text{Dry Soil Weight}} \times 100$$

On separation of 500 gm. dry weight soil block, it was found that sand fraction was 350 gm and that of loam was 100 gm and rest was clay.

Based on data, determine the texture of soil

$$\% \text{ of material present} = \frac{\text{wt of the type of materials (soil)}}{\text{Total wt of soil block}}$$

% of sand = 350/500 100 = 70%

% of loam = 100/500 100 = 20%

% of clay = 50/500 100 = 10%

As from the above given classes, the presence of 70-80% of sand indicates sandy loams.

SOIL TEXTURE

Aim

To determine the texture of a given sample of soil.

Requirements

Soil, Sieve Set, Blotting paper, Balance.

Procedure

A known quantity of soil (40 gm) is taken in the sieve set consists of 3 meshes. The upper most mesh is 250 m. cm size. The middle one is 150 cm in size and the lower one is 125 m cm size. 40 gm of clay soil is taken in the upper mesh and is allowed to pass through all the meshes. The soil present in each mesh is taken separately and weighed. Some quantity of red soil is taken of above procedure and is repeated. Finally the values are noted. From these values, 5% of soil present in each mesh is calculated.

Observation

The soil collected from the uppermesh is comprised of relatively bigger particles and soil in the mesh is weighed. About 24 gm in case of black soil, whereas in case of red soil, it is 16 gm. The size of these particulates is more than 0.2 mm in diameter and is considered as coarse sand. Finally from the lower mesh, 6 gm and 13 gm for black and red soils respectively are obtained and this particle size is less than 0.002 mm.

Discussion

The texture of soil depends upon the size of the particles present in it. The weathered particles are classified on basis of the size. International society of soil sciences, the classification is as follows :

Coarse sand - 2 mm to 0.2 mm

Fine sand - 0.2 mm to 0.02 mm

Silt sand - 0.02 mm to 0.002 mm

Clay sand - < 0.002 mm

Depending upon the proportion of various sizes, soils are classified into sandy, silty, loamy or clay soils. The % may differ from soil to soil.

If 70-80% of sand represents sandy bears 75%, Clay loams; 20 - 30% clay and silt represents 30 - 45% of loam soil.

Result

The given sample of soil is in sandy loam.

EXPERIMENTS ON PHYSICAL CHARACTERS OF SOIL

Particle Size Analysis

Relative proportion of the soil particles of various sizes is an important physical parameter which determines the texture of soil. Large particles help in providing the physical support to the plants, while smaller size particles determine the capacity of soil to hold the water and availability of nutrients.

According to the International system, the range of the particle sizes in the soil is as follows:

Coarse sand - 0.2 - 2.2 mm

Fine sand - 0.2 - 0.02 mm

Silt - 0.02 - 0.002 mm

Clay - < 0.002 mm

The percentages of the above particles can be measured by several methods such as (1) Pipette method (2) Beaker method (3) Hydrometer method and (4) Counter count method.

The most simple and quicker method of all is the Hydrometer method. The method is suitable only when the organic matter, salts and Calcium carbonate are present in low amount. In case of the presence of these substances in appreciable quantities, the pretreatment of the soil is required. The pretreatment of the soil is as follows:

In case of more than 12% of organic matter, it should be removed by Hydrogen peroxide treatment. For this, take 50 g of soil, cover with water and add 20 ml of H_2O_2. Heat until the reaction ceases. If necessary, repeat the process until all the organic matter is destroyed. To remove excess of $CaCO_3$ and add slowly 1 + HCl to 50 g of soil. Warm and stir until reaction subsides. Continue the addition of Hydrochloric acid until excesss acid formed. For saline soils, wash them thoroughly to leach the salts out. Dry the soil and reweight before proceeding.

Apparatus

Standard Hydrometer with Bouyoucos scale in gl. Electrically driven stirrer and Graduated cylinder of one litre capacity.

Reagents

A. Calgon, 5%: Dissolve 50g of Calgon in distilled water, adjust the pH at 9 with Na_2CO_3 and makeup the final volume to 1 litre.

Procedure (for Silt and Clay)

Weigh 50g of air dried 2 mm depth soil for two times and transfer one portion to litre capacity breaker. Keep the second portion of 50g of soil in the oven at 105°C for overnight to findout the moisture contents of it. Add 25 ml of 5% Calgon and 400 ml of water to the breaker soil. Allow the sample to soak for atleast in 10 minutes.

Mix the suspension for thorough dispersion by stirring for 15 minutes with the help of an electrically driven high speed mixer. Transfer the contents to a 1 litre graduated cylinder removing all the adhered soil with the help of a stream of water, and bring the level to the mark of 1 litre with water.

Stir for 1 minute with a paddle or a glass rod. Record the timing for readings with the Soil Hydrometer. Add a drop of Amyl alcohol in case the surface is covered with after stirring. Take the reading according to the following scheme : (Introducing the Hydrometer just 20 sec before reading).

1. 4 min. 48 sec. = International silt and clay (< 20)
2. 5 hours = International clay (< 2)

Take the temperature of suspension after each reading.

Calculation

As normally the Bouyoucase Hydrometer is calibrated for a temperature of 19.5°C, a temperature correction in the readings is required. For temperature correction 0.3 units are either added or substrate for every degree above or below 19.5°C respectively, in the Hydrometer reading.

$$\text{Clay \%} = \frac{A \times 100}{S - M} - 1$$

$$\text{Silt + Clay \%} = \frac{B \times 100}{S - M} - 1 \text{ (1 is substrate as Calgon correction)}$$

$$\text{Silt \%} = \text{(Silt + Clay) \%}$$

Total land % (approx) = 100 (Silt + Clay)%

Where,

A = Hydrometer reading (g/l) after 5 hours.

B = Hydrometer reading (g/l) after 4 min. 48 sec.

S = Weight of soil taken (50 g)

M = Moisture contents in 50g of soil (in g)

WATER HOLDING CAPACITY

Water holding capacity of soil usually refers to amount of maximum water which can be held in the saturated soils. It is generally measured as the amount of water takenup by unit weight of dry soil when immersed in water under standardized conditions.

Requirements

A. Circular Brass Boxes: They are normally of the dimensions of 5.6 cm international diameter and 1.6 cm height. The bottom is performed with several holes (0.75 mm in diameter) spaced at 2 mm. Sometimes, however, instead of circular boxes, square boxes can also be used with more or less same volume.

Procedure

Place a filter paper in the brass box (Whatman No. 1 or 44), of appropriate dimension so as to cover the whole perforated bottom of the brass box. Take the weight of brass box plus filter (W_1). Now transfer the 0.5 mm dry soil in small portion, tapping the box gently after each addition, until the box is nearly full. Then add sufficient soil to fill the box and strike off the extra soil level by moving one spatula over it.

Place the packed brass box in a petridish and add water upto a depth of about 1 cm. After some time, water has been absorbed by the soil, restore the depth of water in petridish by addition of some more water and keep for overnight.

Next day after 12-16 hrs, remove the box, wipe it out dry from outside and take the weight (W_2). Now dry it in an oven for about 24 hrs at 105°C, in a desiccator and again take weight (W_3).

Also findout separately the amount of the water absorbed by the filter paper, for this, weigh a few filter papers (say for example 5 pieces), and after saturating them with water, again take their weight. Calculate the amount of water absorbed by a single filter paper (W_4).

Calculation

W_1 = Weight of brass box + Filter paper

W_2 = Weight of brass box + Saturated soil

W_3 = Weight of brass box + oven - dry soil

W_4 = Amount of water retained by the filter paper

$$\text{Water holding capacity} = \frac{W_2 - W_3 - W_4}{W_3 - W_1} \times 100$$

Temperature

The temperature of soil is more variable at a surface layer, but ceases to show much variation with the greater depth.

Requirements

Soil Thermometer: This actually a mercury thermometer. The build of which is kept inside of a metal cone to facilitate the penetration into the soil. The upper side of a metal cone is fixed with a wooden shaft, which supports the stem of the thermometer to see the reading.

Procedure

Insert the cone in the soil upto a desired depth and note the reading of the thermometer directly on the stem which is open in the air.

pH with pH Paper: For field test, pH of soil is, recorded with the help of universal indicator. Take in different test tubes with a small amount of soil in each, and add to each equal amount of Barium sulphate and about 15 ml of water. Keep the solution for some time till clear supernatant fluid is obtained. Take a few drops of the fluid on a porcelain till and add equal amount of universal indicator. Note the colour procedure and now match this colour with the colour chart for different pH values pasted on the indicator bottle. Read the pH value written on the strip with which the colour of solution resembles or matches.

pH with pH Meter: pH of soil is the measure of "Hydrogen ion concentration" and depends largely on the relative amounts of the absorbed Hydrogen and metallic ions. It is a good measure of the intensity of acidity and alkalinity of a soil-water suspension, and provides a good identification of the soil chemical nature. pH of the soil suspension is highly depends on the soil water cations. Determination of pH at moisture saturation level and 1.5 soil suspension is most common.

Procedure

To determine the pH at the moisture saturation level, take about 50g of soil in a beaker. Add small portions of distilled water, without stirring the soil, until a glistering layer appears at the surface. Now make a uniform paste of soil by stirring with the help of a glass rod.

For determination of pH of soil in 1:5 soil suspension, take 20g of soil and 100ml of distilled water. Stir for about an hour at regular intervals.

Determine the pH electrometrically by using pH meter in unfiltered soil solutions.

Results are expressed directly in pH until associated with the specific dilution of soil suspension, e.g., pH in 1:5 soil suspension.

Conductivity

Conductivity is the measure of current carrying capacity, gives a clear idea of the soluble salts, present in the soil. It is conventionally determined in 1:5 soil suspension. Like water, the conductivity of the soil suspension is also measured by a Conductivity meter.

Apparatus

Conductivity meter, thermometer, glassware to prepare soil suspension.

Procedure

To prepare 1:5 soil solution, take 20g soil in 100 ml of aerated distilled water and shake mechanically for about an hour. Measure the conductivity of the suspension within one hour of preparation on soil suspension by directly dipping the conductivity cell into the suspension. Take the temperature of soil suspension to convert the results at 25°C employing the Table given for Water Analysis.

Calculation and Results

Conductivity at 25°C = Observed conductivity cell constant
Temperature factor.

Conductivity is generally expressed in mmho or mho (a recent unit equilvalent to mho) is Siemens, (S). The results should always be accompanied by the specific dilution of the soil suspension, e.g., conductivity mmho at 25°C in 1:5 soil suspension.

DETERMINATION OF SPECIFIC GRAVITY

Aim

Determine the specific gravity of soil fraction passing 4.75 mm I.S. sieve by density bottle.

Need and Scope

The knowledge of "specific gravity of soil" is needed in calculation of soil properties like void ratio, degree of saturation, etc.

Definition

Specific gravity 'G' is defined as the ratio of the weight of a given volume of soil solids at a given temperature to weight of an equal volume of distilled water at that temperature. Both weights are taken in air.

Planning and Organisation

Apparatus:

1. Density bottle of 50 ml with stopper having capillary hole.
2. Constant temperature bath.
3. Balance to weigh the materials (200 gm capacity) (accuracy 10 mg).
4. Bottle with distilled water.
5. Alcohol and ether.

Test procedure:

1. Clean and dry the density bottle
   (*a*) Wash it with water and allow it to drain.
   (*b*) Wash it with alcohol and drain it to remove water.
   (*c*) Wash it with ether, to remove alcohol and drain ether.
2. Weigh the empty bottle with stopper (W_1) enter in the tabulation.
3. Take about 10 to 20 grams of oven dried soil sample which is cooled in a desiccator. Transfer it to the bottle. Find the weight of the bottle and soil (W_2) enter the tabulation.
4. Put 10 ml of distilled water in the bottle to allow the soil to soak completely. Leave it for about 2 hours.
5. Again fill the bottle completely with distilled water put the stopper and keep the bottle under constant temperature water baths (T_x °C).
6. Take the bottle outside and wipe it clean and dry note. Now determine the weight of the bottle and the contents. (W_3) enter the tabulation.
7. Now empty the bottle and thoroughly clean it. Fill the bottle with only distilled water and weigh it. Let it be W_4 at temperature (T_x °C).
8. Repeat the same process for 2 to 3 times, to take the average reading of it.

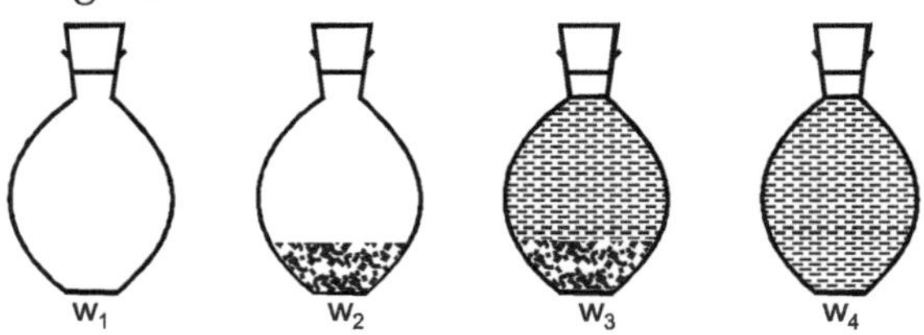

Fig. 16.3 : *Specific gravity botles*

Observations

S.No.	Observation number	1	2	3
1.	Weight of density bottle (W_1)g	21.78	20.98	21.60
2.	Weight of density bottle + dry soil (W_2)g	38.49	37.52	39.85
3.	Weight of bottle + dry soil + water at temperature T_x°C (W_3)g	79.81	79.52	82.40
4.	Weight of bottle + water (W_4)g at temperature T_x°C	69.32	69.11	70.95
	Specific gravity 'G' at T_x°C	2.686	2.698	2.684
	Average specific gravity 'G' (at T_x°C)		2.689	

Calculation

The specific gravity of the soil is calculated as:

$$= \frac{\text{Dry weight of soil}}{\text{Weight of water of equal volume}}$$

$$= \frac{W_2 - W_1}{(W_4 - W_1) - (W_3 - W_2)}$$

$$= \frac{W_2 - W_1}{(W_2 - W_1) - (W_3 - W_4)}$$

$$= \frac{W_d}{W_d - (W_3 - W_4)}$$

Interpretation and Reporting

Unless or otherwise specified specific gravity values reported shall be based on water at 27°C. So the specific gravity at 27°C = K Sp. gravity at T_x°C.

$$\text{Where } K = \frac{\text{Density of water at temperature } T_x\,^\circ\text{C}}{\text{Density of water at } 27^\circ\text{C}}$$

The specific gravity of the soil particles lie within the range of 2.65. Soils containing organic matter and porous particles, may have specific gravity values below 2.0. Soils having heavy substances may have values above 3.0.

BULK DENSITY AND DRY DENSITY (Field Density Test)

Aim

To determine the bulk and dry densities of a sample obtained in core cutter of a compacted specimen.

Need and Scope

The *in situ* density of a soil is to be known before the bearing capacity or settlement characteristics are worked out. Similarly in case of earth dams and embankments, whether the soil has been compacted to a specified value of dry density or not has to be known for further continuation of work.

Distinctly, two different unit weights or densities are often referred to, *viz*. (*i*) The total or bulk unit weight γ_1 or which is the ratio of the weight of soil including water to the total volume which includes the air volume as well; (*ii*) The dry unit weight which is the ratio of the solids to the total volume i.e. volume of solids + volume of air, as water is not present.

Planning and Organisation

Apparatus: The following equipments are necessary to determine the insitu density of soil using the core-cutter method :

1. Sampling tube 38 mm internal diameter and of length between 100 and 120 mm.
2. Wooden cap and wooden mallet.
3. A spade to remove soil upto the sampling depth.
4. A sample mould to trim the soil sample with spatula.
5. Hydraulic sample ejector; alternatively, a screw type.
6. Weighing balance, moisture cans, electric oven thermostatically controlled as specified in water content test.

Procedure

(*a*) Top soil is removed in the field to the depth at about which the *in situ* density is required.

(*b*) Atleast 3 sampling tubes are driven, not very close to one another, by placing the wooden cap and hammering mallet.

(*c*) The soil around the tubes is dug out and the samples are removed.

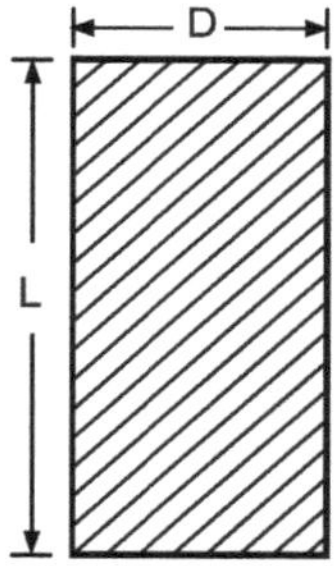

Fig. 16.4: *Core cutter soil sample*

(*d*) The soil is then ejected from the tubes by using either a hydraulic or a screw type ejector.

(*e*) The sample is then placed with in a split mould and the edges trimmed off; Edge trimmings in contact with the soil specimen is taken for the water content determination.

(*f*) The trimmed soil is cylindrical with a diameter of 38 mm and length of 76 mm. This is weighed to a suitable accuracy (W_1). The bulk density of the soil specimen is determined from the expression.

$$\gamma_b = \text{Wb/Sample volume g/cm}^3.$$

(*g*) The water content is determined after finding the oven dried weight of the sample trimmings that had been kept in the oven for 24 hours. The dry density of the insitu soil is,

$$\gamma_d = \frac{\gamma_b}{(1+w)}$$

where 'w' is the water content expressed in decimals.

Precautions

1. This method is possible where the soil contains some percentage of clay.
2. Thin film of lubrication oil is to be applied to outer and inner surface of tube so that from the tube and sample can easily be ejected.

Observations

S.No.	Dia cm	Length cm	Volume cm^3	Weight gms	Bulk gm/cm^3 density	Dry wt gms	Water % content	Dry density
1.	3.81	7.62	86.79	125.84	1.44	109.20	15.23	1.24
2.	3.81	7.62	86.79	132.79	1.53	116.30	14.17	1.34
3.	3.81	7.62	86.79	126.68	1.45	109.38	15.81	1.25

SAND REPLACEMENT METHOD (Field Density Test)

Aim

Determine the insitu density (in g/cm^3 or kg/cm^3) of natural or compacted soils using sand pouring cylinders (Fig. 16.5).

Need and Scope

The insitu density of natural soil is needed for the determination of bearing capacity of soils, for the purpose of stability analysis of slopes, for the determination of pressures on underlying strata of calculational settlement and the design of underground structures.

It is a very important quality control test, where compaction is required, in the cases like embankment, earthdam and pavement construction.

Planning and Organisation

Apparatus:

1. Sand pouring cylinder of 3 lits/16.5 lits capacity, mounted above a pouring cone and separated by a shutter cover plate (Fig. 15).
2. Tools for excavating holes; suitable tools, such as scraper tool to make a level surface.
3. Cylindrical calibrating container with an internal diameter of 100 mm/200 mm and an internal depth of 150 mm/250 mm fitted with a flange 50 mm/75 mm wide and about 5 mm surrounding the open end.
4. Balance to weigh upto an accuracy of 1 g.
5. Metal containers to collect excavated soil.
6. Metal tray with hole : 300 mm/450 mm square and 40 mm/50 mm deep with a 100 mm/200 mm diameter hole in the centre.
7. Glass plate about 450 mm/600 mm square and 10 mm thick.
8. Clean, uniformly graded natural sand passing through 1.00 mm I.S. sieve and retained on the 600 micron I.S. sieve it shall be free from organic matter and shall have been oven dried and exposed the atmospheric humidity.
9. Suitable non-corrodible air-tight containers.

10. Thermostatically controlled oven with interior on noncorroding material to maintain the temperature between 105°C to 110°C.
11. A desiccator with any desiccating agent other than Sulphuric acid.

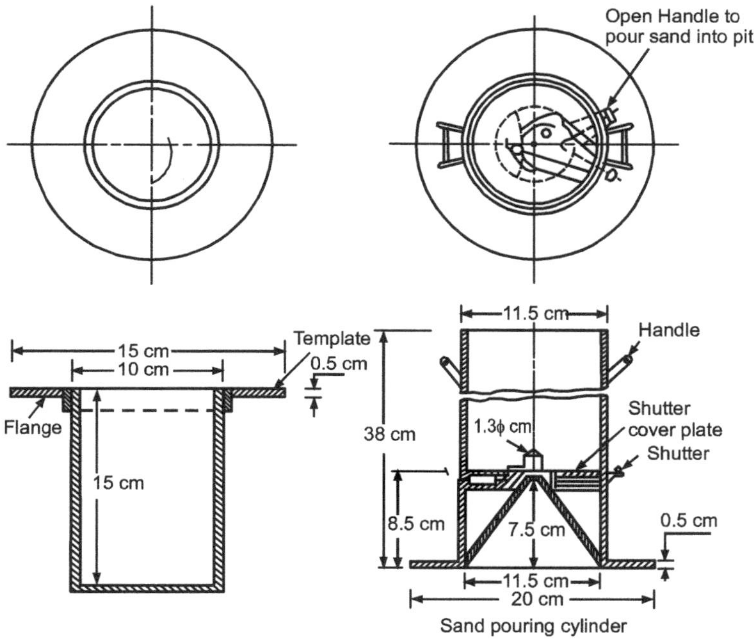

Fig. 16.5 : *Field density apparatus*

Theory

By conducting this test, it is possible to determine the field density of the soil. The moisture content is likely to vary from time to time and hence the field density also. So it is required to report the test result in terms of dry density. The relationship that can be established between the dry density with known moisture content is as follows :

$$\gamma_d = \frac{\gamma_b}{1+w}$$

γ_d = Dry density

γ_d = Bulk density

w = Water content

Procedure

Calibration of the Cylinder

1. Fill the sand pouring cylinder with clean sand so that the level of the sand in the cylinder (Fig. 16.5) is within about 10mm from the top. Findout the initial weight of the cylinder plus sand (W_1) and this weight should be maintained constant throughout the test for which the calibration is used.

2. Allow the sand of volume equal to that of the calibrating container to run out of the cylinder by opening the shutter, close the shutter and place the cylinder on the glass plate. Open the shutter and allow the sand to run out. When no faster movement of sand takes place in the cylinder close the shutter and remove the cylinder carefully. Weigh the sand collected on the glass plate. Its weight (W_2) gives the weight of sand filling the cone portion of the sand pouring cylinder.

   Repeat this step atleast three times and take the mean weight (W_2).

   Put the sand back into the sand pouring cylinder to have the same initial constant weight (W_1).

Determination of Bulk Density of Soil

3. Determine the volume (V) of the calibrating container by filling it with water to the brim. Check this volume by calculating from the measured internal dimensions of the container.

4. Place the sand pouring cylinder centrally on the top of the calibrating container making sure that constant weight (W_1) is maintained. Open the shutter and permit the same to run into the container. When no further movement of sand is seen close the shutter, remove the pouring cylinder and find its weight (W_3). But the sand into the sand pouring cylinder (Fig. 15).

Determination of Dry Density of Soil

5. Approximately 60 sq cm of area of soil to be tested should be trimmed down to a level surface (approximately of the size of the container). Keep the metal tray on the level

surface and excavate a circular hole of volume equal to that of the calibrating container. Collect all the excavated soil in the tray and findout the weight of the excavated soil (W_ω). Remove the tray, and place the sand pouring cylinder filled to constant weight (W_1) so that the base of the cylinder covers the hole concentrically. Open the shutter and permit the sand to run into the hole. Close the shutter when no further movement of the sand is seen. Remove the cylinder and determine its weight (W_3).

Note: It is necessary to make a number of repeated determinations (atleast 3) to average the result. Since the dry density varies from point to point, the number of determinations should be such that an additional one would make no significant difference to the average.

6. Keep a representative sample of the excavated soil for water content determination.

Observation, Results and Calculations

The test results may be tabulated as given below using the appropriate symbols and words in each case.

Name of the Project : ....

Tested by : ....

Location Area : Other particulars (Temp, Rainy etc)

S.No.	*Sample Details Calibration*	*1*	*2*	*3*
1.	Mean weight of sand in cone (of pouring cylinder) (W_2) in gms	424	422	423
2.	Volume of calibrating container (V) in ml	1010	1010	1010
3.	Weight of sand (+ cylinder) before pouring (W_3) in gms	8802	8800	8804
4.	Mean weight of sand (+ cylinder) after pouring (W_3) in gms	6985	6965	6972
5.	Weight of sand to fill calibrating ($W_a = W_1 – W_3 – W_2$) in gms	1393	1413	1409
6.	Bulk density of sand γ_s = Wa/V gms/cc	1.379	1.399	1.395

S.No.	Measurement of Soil Density	1	2	3
1.	Determination No. 1, 2, 3..........			
2.	Weight of wet soil from hole ($W\omega$) in grms	1930	1912	1292
3.	Weight of sand (+ cylinder) before pouring (W_1) in gms	8802	8804	8804
4.	Weight of sand (+ cylinder) after pouring (W_4) in gms	6981	6950	6969
5.	Weight of sand in hole ($W_b = W_1 - W_4 - W_2$) in gms	1452	1428	1417
6.	Bulk density $\gamma_b = (W_\omega / W_b) \times \gamma_s$ gm s/cc	1.859	1.869	1.880
7.	Water content container No.			
8.	Weight of soil for water content determination in gms	168	165	167
9.	Weight of oven dried soil in gms	146	144	145
10.	Water content 'ω'%	15.07	14.58	15.17
11.	Dry density $\gamma_d = \dfrac{\gamma b}{1+\omega/100}$ gms/cc	1.616	1.631	1.632

General Remarks

1. While calibrating the bulk density of sand great care has to be taken.
2. The excavated hole must be equal to the volume of the calibrating container.

EXPERIMENTS ON CHEMICAL CHARACTERS OF SOIL

Calcium and Magnesium

Cations present in the exchange complex of the soil can be removed by leaching the soil with Ammonium acetate solution. Different exchangeable cations are then estimated separately in this Ammonium acetate leachate.

For determination of the exchangeable cations, first the soil washed with Ethyl alcohol to remove the soluble fraction. However, for total cations (exchangeable + soluble), there is no necessity of prior alcohol washing. For soluble fraction only, 1:5 soil solution is used to estimate the cations.

Reagents

A. Ethyl alcohol, 40%

Mix 600 ml of distilled water with 400 ml of absolute or 95% alcohol.

B. Absolute alcohol.

C. Ammonium acetate solution.

Dilute 57 ml of glacial Acetic acid to 800 ml with distilled water and the neutralize to pH 7.0 with concentrated NH_4OH. Makeup the final volume to one litre.

D. Aquaregia

Mix 3 parts of Conc. HCl with 1 part of Conc. HNO_3.

E. All other reagents used in determination of Calcium and Magnesium in water.

Procedure

Take 50g of air dried soil in a 500 ml beaker and add about 100 ml of 40% alcohol. Shake well and keep for about 15 minutes. Filter the suspension through Whatman No. 50 filter paper using Buchner funnel and Vacuum pump. Wash the soil 4 or 5 times with 50 ml portions of 40% alcohol and perform the final washing with absolute alcohol to dry the soil. Remove the filter paper and scrap the soil in a 25 ml beaker and wash finally the Buchner funnel and filter paper with 100 ml Ammonium acetate solution to remove and adhered portion of the soil. Stir the soil suspension and keep for overnight. Now filter the supernatent and later the soil with additional Ammonium acetate through Whatman. No. 42 filter paper, using Buchner funnel. Leach the soil 4-5 times more with portions of Ammonium acetate and makeup the final volume of the filtrate to 500 ml with distilled water in a volumetric flask. Ammonium acetate and dispersed organic matter when present in appreciable amounts, interfere with the titration with EDTA. They must almost entirely be removed prior to titration with EDTA. Evaporate a portion of the Ammonium acetate extract to dryness and dissolve the residue in a very small portion of aquaregia. Evaporate again to dryness and dissolve the residue this time in distilled water makeup the original volume of the extract evaporated.

Findout the concentration of Calcium and Magnesium following the method described for the analysis of water.

In case total cations (exchangeable + soluble) or when soluble fraction is very less, do not apply the alcohol washing and add directly 100 ml of Ammonium acetate to 50 g of soil and proceed further.

Calculation

$$\% \text{ Calcium} = \frac{A \times 400.8 \times V}{V \times 1000 \times S}$$

$$\text{Calcium, meq/100g} = \frac{A \times 400.8 \times V}{V \times 20.04 \times 10 \times S}$$

$$\% \text{ Magnesium} = \frac{(B - A) \times 400.1 \times V}{v \times 1000 \times S \times 1.645}$$

$$\text{Magnesium. meq/100 g} = \frac{(B - A) \times 400.1 \times V}{v \times 10 \times S \times 1.645 \times 19.16}$$

Where

A = Volume of EDTA (ml) used for Calcium determination

B = Volume of EDTA (ml) used for Calcium + Magnesium determination.

V = Total volume of soil extract prepared (in this case 500 ml)

S = Weight of soil taken (in this case 50 g).

v = Volume of soil extract titrated (ml).

Values can also be converted in mg/100g (mg per cent after multiplying the % values by 1000).

Sodium and Potassium

Exchangeable Sodium and Potassium can also be determined in Ammonium acetate leachate by Flame photometric method.

Procedure

Use the remaining Ammonium acetate leachate of soil after Calcium and Magnesium determination for Sodium and Potassium. Findout the Na and K by Flame photometry as described in case of Water Analysis.

Calculation

$$\% \text{ Na} = \frac{mg\ Na / l\ of\ soil\ extract \times V}{1000 \times 4}$$

$$\text{Na meq/10g} = \frac{mg\ Na / l\ of\ soil\ extract \times V}{1000 \times S}$$

$$\% \text{ K} = \frac{mg\ K / l\ of\ soil\ extract \times V}{1000 \times S}$$

$$\text{K mgq/100g} = \frac{mg\ K / l\ of\ soil\ extract \times V}{1000 \times S \times 39}$$

Whereas

V = Total volume of soil extract prepared

S = Weight of soil taken

To convert the values in mg/100g (mg per cent) multiply the % values by 1000.

Phosphorus

Phosphorus in soil is determined as available Phosphorus, which can be extracted from soil with 0.002 (NH_2 SO_4) (1 Soil : 200 H_2SO_4). It is always advisable to measure P in fresh soil, whatever, if the soils are to be preserved in air, dry them instead of oven drying to minimise the error.

Reagents

(A) Sulphuric acid, 0.002 N: Prepare a stock solution of 0.1 NH_2SO_4 by diluting conc. H_2SO_4 360 times (2.78-1000 ml) and standardize with an alkali. A desired volume of stock solution is diluted 50 times to obtain 0.002 NH_2SO_4. Add 3g of Ammonium sulphate $(NH_4)_2$ SO_4 or Potassium sulphate, K_2SO_4 per litre to produce a final pH of 3.0

(B) All other reagents used in determination of inorganic phosphorus in water.

Procedure

Take a 2 mm fresh soil and determine its moisture content. The moisture content is the difference in weight between fresh and after its drying in oven at 105°C. Now take fresh soil equivalent to

1.0 g oven dry weight in a 500 ml conical flask and add 200 ml 0.002 NH_2SO_4. In case of air-dried soil, directly 1.0 g of soil is taken. Shake the suspension atleast for 1/2 hour. Filter through Whatmen No. 50 filter paper to get a clear solution. Findout the concentration of Phosphorus following the method given for determination of inorganic P in water.

Calculation

$$\% \text{ available } P = \frac{mg\ P / l\ in\ soil\ solution}{50}$$

$$P\ mg\ /100\ \text{g} = \frac{mg\ P / l\ in\ soil\ extract \times 1000}{50}$$

Nitrogen

Nitrogen in the soil is present mainly in the organic form together with small quantities of ammonium and nitrate forms. The most common method for finding out Nitrogen in soil is Kjeldahl method, which measures only organic and Ammonium forms excluding Nitrate.

Apparatus and Reagents

A. Distillation assembly

B. Sulphuric acid

H_2SO_4 Conc. (Sp. gr. 84)

C. Hydrochloric acid 0.1 N

(See determination of alkalinity in Water Analysis)

D. Digestion catalyst

Grind together 20g Copper sulphate, 3g Mercuric oxide (HgO) and 1 g of Selenium powder. Mix thoroughly 1 part of this mixture with 20 parts of Sodium sulphate ($Na_2\ SO_4$) or Potassium sulphate (K_2SO_4).

E. Sodium hydroxide, 40%

Dissolve 40g of NaOH in 100ml of distilled water.

F. Boric acid + mixed indicator.

See determination of Ammonia in Water Analysis.

Organic Matter

Reagents

A. Potassium dichromate, 1 N

Dissolve 49.040 g of $K_2Cr_2O_7$ solution.

B. Suphuric acid, concentrated.

Take H_2SO_4 (sp. gr. 1.84) and add 1.25g Silver sulphate to each 100 ml of acid.

C. Diphenylamine indicator

Dissolve 0.5g diphenylamine in a mixture of 10ml conc. H_2SO_4 and 20 ml distilled water

D. Phosphoric acid

H_2SO_4, 85% (sp. gr. 1.71)

Procedure

Weigh suitable quantity of 0.5 mm perfectly dried soil not exceeding 10 g (containing about 10-25 mg carbon) and transfer it to a dried 500 ml conical flask. Add 10ml of 1N. $K_2Cr_2O_7$ and 20 ml concentrated Sulphuric acid having Silver sulphate dissolved in it, and mix by gentle swirling. Allow to stand the flask for about 30 minutes and after when reaction is over Dilute the contents by 200 ml of Phosphoric acid and 1 ml diphenylamine indicator. The colour will change to bluish purple. Titrate the contents with Ferrous ammonium sulphate, carefully, until the blue colour changes to brilliant green. The end point is very sharp in this titration. If more than 8 ml of the 10ml added $K_2Cr_2O_7$ is consumed, (titration value $<$ 2 ml). Repeat the procedure with less quantity of the soil. If before the titration only, the colour becomes green, repeat with smaller quantity of soil again.

Calculation

$$\% \text{ Carbon} = \frac{V1 - V2}{W} \times 0.003 \times 100$$

$$\% \text{ Organic matter} = \% \text{ C} \quad 1.724$$

Where

V1 = Volume of $K_2Cr_2O_7$ (10 ml)

V2 = Volume of Ferrous Ammonium sulphate

W = Weight of the soil taken.

Depending on the soil only 60-90% of the total organic matter is recovered in this method. No recovery factor has been taken into account in the formula given. Therefore the results should be accompanied by with a mention by "Walk and Black method".

Calcium Carbonate

Rapid titration method

Reagents

A. Hydrochloric acid, 1 N

   Dilute 175 ml of concentrated HCl to 2 liters

   Standardization is not required in this case.

B. Sodium Hydroxide, 1 N

   Dissolve slightly more than 40 g of NaOH in 1 litre of distilled water and standardize with a standard acid. Dilute to the exact strength.

C. Brom Thymol blue.

   Weigh 0.1 g of Brom Thymol blue and dissolve in 1.6 ml of 0.1 N NaOH by grinding in an agate mortar, and dilute to 250 ml.

Procedure

Take exactly 5 g of dried soil and transfer to 150 ml beaker. Add exactly 100 ml of 1 N HCl preferably with a pipette, and cover with a watch glass. Allow to stand for atleast 1 hour with vigorous string intermittently. Allow to settle finally, and remove 20 ml of supernatant with the help of a pipette in a conical flask. Add 6-8 drops of Brom Thymol blue indicator and titrate with 1 N NaOHK at the end point, the colour changes to blue.

Carryout a blank determination by using the same chemicals but without soil.

Calculation

$$\% CaCO_3 - (B\text{-}T) \quad 5$$

where,

B = reading with blank

T = reading with soil

Carbonate Content

Take a little amount of each of the different types of soil in different test tubes. Add few drops of HCI to each tube and note the degree of effervescence in each. The degrees of effervescence produced in all cases may be divided into different groups, as +, ++, +++, ++++, or A, B, C, D. The sample showing maximum amount of Carbonate content present.

Observations

Record the data as shown below :

Sol sample	*Degree of effervescence seen*
1	+ or A
2	++++ or D
3	++ or B
4	+++ or C

Boron

Boron is an essential minor nutrient for plant growth. The tolerance range between toxicity and deficiency is quite narrow and for this reason it may be a pollution hazard at a slightly elevated concentrations.

Precaution should be taken to avoid Boron silicate glassware in the intial steps of the procedure (except for colour development).

Reagents

A. Carmine solution, 0.1 %

Dissolve 1 g of Carminic acid in 1 litre of Sulphuric acid. (Sp. gr. 1.84).

It may take about 2 hours to complete dissoution.

B. Concentrated Sulphuric acid

H_2SO_4 (A.R; Sp. gr. 1.84)

C. Concentrated Hydrochloric acid

HCl (A.R.) 12 N

D. Standard Boron solution (100 mg/ml)

Dissolve 0.5716 g. of Boric acid (H_2BO_3) in distilled water to make up 1 litre of solution. This solution contains 100 mg B per litre.

Procedure

Take 10 g of air-dry soil in a soda glass flask, and boil with 50 ml distilled water for 5 minutes. Filter through a Whatman No. 44 filter paper and makeup the final volume to 100ml.

Find out the concentration of Boron following the method given for Boron determination in Water Analysis.

Calculation

$$\text{Boron, mg/100g} = \frac{mg\ B\ /\ l\ of\ soil\ extract \times V}{S \times 10}$$

Where

V = Total volume of soil extract prepared

S = Weight of the soil taken

SOIL MAPS (Courtesy : Soil Survey Manual, USDA, 1993)

Imagery to Air Field Operations

Aerial photographs are used as the mapping base in most soil survey areas in the United States today. With few exceptions aerial photographs are by far the most practical mapping base for field use by soil scientists. Several kinds of aerial photography is sensitive to approximately the visible portion of the electromagnetic spectrum (wavelengths of 0.38 to 0.78 micrometer). Colour photography covers a similar range. Infrared photography, which covers radiation of somewhat longer wavelengths, is also available. The main kinds of aerial photography are described as follows :

Single-lens aerial photography

The two basic types of aerial photographs are vertical and oblique. Single-lens vertical photographs are the best for soil mapping, although oblique or multiple-lens photographs can be used when rectified. USDA specifications for single-lens aerial photography require an overlap in line of flight of about 60 per cent and a sidelap between adjacent flight lines of an average of 30 per cent. With this overlap, all ground images appear on two or more photographs exposed from different air positions, providing stereographic coverage. Two consecutive photographs within a line of flight are called a stereographic pair.

If every other photograph in a continuous line of flight is removed the remaining photographs provide alternate coverage. Adjoining photographs of alternate coverage in the same line of flight are called alternate pairs. Alternative pairs overlap about 20 per cent too little of permit stereoscopic study of the entire area. Using alternate coverage, instead of full stereographic coverage for mapping, leads to problems with relief displacement during map compilation. Alternate coverage is inadequate for constructing maps by photogrammetric methods based on complete stereographic coverage.

Photographs are exposed on film at a predetermined scale and fixed negative size. The scale depends on purpose. Most USDA aerial photographs are taken with a 153 millimeter lens. (Scale ranges from 1:38,000 to 1:80,000). Satisfactory enlargements upto 1:7,920 can be made from 1:40,000 negatives. Most aerial cameras currently in use by expose an image of about 23 by 23 centimeters.

Photographs made directly from the original negatives at the same scale are called contact prints. In contact printing, errors cannot be rectified and the scale cannot be changed. Contact prints are economical to make and have better resolution than enlargements.

Photographs can be readily enlarged or reduced; this is one of their advantages. The process is slower and more expensive than contact printing. Some details are lost in the preparation of enlargements, but the loss is small when skilled operators use modern processing equipment and the original negatives.

Enlarging has certain advantages. All prints for an area can be brought to a nearly uniform scale. Tilt, which causes displacement of objects and scale distortion, can be rectified. Such operations require more time than simple enlarging, but later savings may more than offset the cost of bringing photographs to a common scale. If the photograph is enlarged more than 5 times, prints are usually unsatisfactory. Enlargement increases the size of the photograph as well as the scale. The size of sheet varies with the enlargement. If the contact print at a scale of 1:20,000 is 23 cm square, an enlargement to 1:15,840 will be 29 cms square and an enlargement to 1:7,920 will be 58 cm square.

Photo indexes are inexpensive and should be obtained when available. They are useful for determining the number and location of individual photographs within an area. They are also useful for schematic mapping and for preliminary studies.

The greatest advantage of aerial photography in soil surveying is the wealth of ground details be shown. Field boundaries, isolated trees, small clumps of bushes, rock outcrops, and buildings are visible and assist in orienting the mapper and in plotting the soil boundaries and other features. Both the speed and accuracy of the work are increased by using photographs. Base maps for publication can be constructed from aerial pholographs economically and in a reasonable time. Showing all of the intricate cultural and physical details, a stereographic series provides a relief model of the area.

Aerial photographs also have some disadvantages and limitations in soil surveying. Elevations are not shown. Scale is not precisely uniform. Differences of scale between adjoining photographs create some minor difficulties in matching and transferring soil boundaries. Distances and directions cannot be measured as accurately as on topographic maps or some other kinds of photographs because of distortions caused by tilt, image displacement, and other inherent errors. Finally, although far more detail is shown on aerial photographs than on most maps, the detail is not always as legible and more skill is required to interpret the photograph. Nevertheless, the advantages of aerial photographs generally greatly outweigh the limitations.

Photographic indexes are available for most of the photography available from Federal agencies. Indexes are prepared by fastening together the individual prints of an area. The images are matched, and the photographs are overlapped so that all marginal data are visible. The assembly is then photographed at a smaller scale. Most indexes prepared by the United States Department of Agriculture have a scale of 1:63,360 or 1:126,720.

Once a survey has been scheduled, photographs should be ordered as soon as possible. The order gives the exact boundaries of the survey area, the scale of photography needed, desired coverage (stereographic or alternate), and the date that fieldwork is to begin. Any special requirements, such as weight or finish or paper, are stated. Low-shrink paper is recommended for most field-mapping.

Panchromatic photography records all colours in varying shades of gray. Most modern black and white photography is of excellent quality. Because of their quality and economy, photographs made from panchromatic film are the most widely used for soil surveys.

Colour photography records features of the surface in colours of the visible spectrum. The colours on the print are about the same as the colours of the features when the photograph was taken, but the colours of the ground features may be different at other times. The colour of a soil also may differ, according to such factors as Sun angle, atmospheric conditions, delays between flights, and moisture state of the surface. The cost for obtaining colour photography is about 1½ to 2 times as much as panchromatic photography. Colour prints cost 2½ to 4 times as much as black and white prints. Excellent black and white prints can be made directly from colour negatives at the same cost as prints from a panchromatic film.

With high-altitude photography, fewer photographs are required to cover an area. Contact prints of the original nagatives can be used for stereographic coverage. Enlarged stereographic coverage can be prepared from selected stereographic pairs. Special stereoscopes are helpful when viewing the larger prints in field offices.

For some soil surveys, photobase maps are printed from high-altitude photography and low-shrink paper. In other surveys the photobase maps are printed on transparent film with a matty surface. Normally, field mapping on outdated photographs is transferred to film prints, and paper prints are used for new field mapping.

Obtaining high-altitude photography and preparing photobase maps nearly always cost less than constructing photobase maps from a controlled aerial mosaic. High altitude photographs have better image quality than controlled mosaic ones.

Infrared photographs records a portion of the spectrum that is not visible to the human eye. Infrared film is also sensitive to part of the visible spectrum, but true infrared photography is exposed through a deep red filter so that only the infrared radiation is recorded. Prints from infrared film have distorted shades of gray in comparison to prints from panchromatic film. Bodies of water and areas in shadow appear black. Broad leaved trees appear very light, as though covered with frost. Foliage of coniferous trees appears distinctly darker. Roads are dark, instead of very light as on panchromatic prints. These characteristics are useful for detecting patterns of soil moisture states, identifying forest types, and detecting vegetation under stress from disease or other causes. Infrared aerial photography is especially valuable in areas having atmospheric haze because the film is not sensitive to the blue portion of the spectrum that it normally associated with haze. Infrared photography costs about 10 per cent more than panchromatic photography.

Modified infrared photography is a compromise between true infrared photography and panchromatic photography. The images have some of the characteristics of each. At first glance, a modified infrared photography looks very much like a photograph made from panchromatic film. It shows more contrast between some kinds of vegetation and records differences in soil wetness in more distinctive patterns than panchromatic photographs. Modified infrared photography costs about 10 per cent more than panchromatic photography. Prints from infrared negatives cost the same as prints from panchromatic film.

Colour infrared photography is sensitive to the green, red and infrared portions of the electromagnetic spectrum. It produces false colours for most objects. The prints are spectacular; the colours are often brilliant and contrasting. This type of photography is especially useful for the study of vegetation. Vigorously growing vegetation appears brilliant red. Colour infrared photography costs about the same as Conventional colour photography.

REMOTE SENSING

Remote Sensing refers to the full range of activities that collects information from a distance. It includes photography, which has been the most widely used remote sensing technique for many years. The range of the electromagnetic spectrum that can be sensed from a distance, however, is much greater than that covered by conventional photography. Other techniques have been devised to use part of this range. (*Remote sensing: "With special reference to Agriculture and Forestry". Committee on Remote Sensing for Agricultural purposes, Agricultural Board, National Research Council, Washington, D.C., National Academy of Sciences, 1970*). Non photographic sensors can perceive the parts of the electromagnetic spectrum from ultraviolet wavelength of 100 cm.

The extent to which some of the newer Remote sensing techniques can be used in soil surveys has not been fully explored. Field work cannot be eliminated, but how much it can be reduced is not clear. Soils must be examined to a depth of about 2 m or to solid rock—beyond the present reach of most remote sensors or combination of sensors. Atleast some clues to many soil properties are provided by surface features. It is these clues, many of them quite subtle and obscure, that are sought and used in drawing soil boundaries. These clues also assist the making of accurate soil maps without excessive digging or probing. Remote sensoring contributes greatly to soil surveys by revealing these clues. The imagery extends hard data about soils and their formation to new areas.

In areas of the country where it can be used, Ground penetrating radar (GPR) and statistical analysis of the radar data can be useful aid to soil mapping and can provide an effective and efficient method to characterize variability within soil map units. GPR has the advantage of observing a linear transect of the soil continuum across a landscape.

The prospect of using more than one set of imagery is important. Such a set might be made up of two or more kinds of photography made at the same time multiband photography or two sets of one kind of photography made at different times of the year, or some combination of these. Although several sets of photographs and other imagery are likely to yield more clues about soils than one set, the extra cost for the additional clues would have to be justified.

Photograph like images can be made by non photographic sensors of any part of the electromagnetic spectrum. Hence, outputs from the sensors can be viewed and treated like photographs. An example is *side looking radar,* which can penetrate clouds and can be used at night as well as in the daytime. The radar can produce prints that resemble photographs, although the images are not as clear as panchromatic photographs. Side looking radar is useful where continuing cloud cover prevents conventional photography. For use with computers, impulses from side looking radar and other non photographic sensors can go directly into automatic data processing systems for storage of analysis.

Space exploration has added a new dimension to Remote sensing. Earth orbiting satellites can be equipped with several kinds of sensors, including cameras. Imaging from space has the same problems as imaging from aircraft and the additional problem of transmitting the data to earth. Imaging from space has two important advantages. First, large areas – thousands of square kilometers – can be examined from a single point in orbit. Second, any area can be repeatedly examined on a regular schedule.

Base Material

More than one kind of cartographic material suitable as a mapping base could be available for an area. The choice of base material depends on the relative advantages of available material for all aspects of the job including map compilation and reproduction as well as fieldwork.

Selecting the mapping base: The quality of the cartographic material used in mapping and for publication affects the accuracy of map unit boundaries and soil identification, the rate of progress, the methods and costs of map construction, and the quality of the published map. The assembly of cartographic materials should begin as soon as an area is selected for survey.

For most surveys, purchasing new or recent photography and preparing field sheets at the dimension and scale that will be used for publication is an economically sound practice. Some of the costly steps of map compilation are eliminated. High altitude aerial photographs are particularly suitable, as are orthophotographs. Such photography is precise enough to eliminate the preparation of a costly controlled mosaic.

Plans for the survey must consider all costs of map construction—fieldwork, copulations, finishing, and publication. Plans must be made in an advance of field operations, especially if contracts are to be let for new photography. Completion of aerial photography contracts can be delayed for a long time by adverse weather conditions.

In ordering new photography, time must be allowed for preparing specifications, awarding contracts, photographing the area and inspecting before accepting the work. The cost of original aerial photography varies greatly.

Enabling aerial photography contractors to keep their equipment and personnel busy throughout the year and taking advantage of favourable seasonal conditions reduces the cost of aerial photography. Such factors as geographic latitude and solar altitude must be considered in scheduling flights to reduce or eliminate objectionable shadows. Trees should be bare and other vegetation at a minimum for the best results. This requirement further limits the flying season in the northern half of the United States. Moisture conditions are important in revealing soil patterns. In areas of the central United States where annual row crops are the main type of crop, the lack of ground cover and soil moisture conditions are nearly optimum for indicating soil pattern sometime between late April and the end of June. For economy, scheduling requires close study of regional weather patterns in a survey area in order to forecast the number of "photographic days" (no more than 10 per cent cloud cover) in each month.

Orthophotographs: An orthophotograph is an aerial photograph with nearly all the image displacement and scale errors removed. Aerial photographs are converted to orthophotographs by simple rectification for low-relief terrain or by differential rectification for high-relief terrain. Orthophotography is prepared by methods designed to meet National Map Accuracy Standards. Various accuracy tests are performed to verify that 90 per cent of the well defined points tested are within 12.19 meters of true horizontal position, the horizontal accuracy standards for a 1:24,000 scale. An *orthophotoquad* is an orthophotograph formatted to the same size and scale as any of the USGS topographic quadrangles.

Orthophotographs portray is an abundance of details and have correct scale and positional accuracy that is not found in conventional aerial photography. Production costs of Orthophotographs compare favorably with controlled mosaic production costs. Orthophotographs of varying scales are used as base maps for soil surveys, land use planning, resource studies, and topographic maps. Orthophotography can be enhanced with such cartographic features as contours, political boundaries, highways, and principal places to provide maps designed to meet the general need of most users.

Aerial mosaics: Aerial mosaics are made by matching and assembling individual photographs to form a continuous image of an area. Several methods of assembly are used, and the resulting mosaics vary widely in accuracy and usefulness.

The two general types of aerial mosaics are uncontrolled and controlled. An *uncontrolled mosaic* is made by simple matching like images on adjoining photographs without geographic control of the positions of the features. A *controlled mosaic,* displays photographs that are very close to uniform scale and rectified to reduce tilt and displacement. Features on the mosaic are close to their correct positions on the map grid. The accuracy of a controlled mosaic approaches that of a good planimetric map.

Between the uncontrolled mosaic covers a larger area than a single photograph, fewer photobase sheets need be matched and the chance for error is reduced. A mosaic can be made to cover a specific area, such as a township or a drainage basin.

Topographic maps: Topographic maps are not photographs. A topographic map represents horizontal and vertical positions of physical features by using standard symbols. Published maps usually show cultural features such as roads, railroads, and buildings in black, drainage features in blue, and contour lines in brown. Some also show additional features, such as vegetation in overprints of green or other colours.

Most topographic maps published by the U.S. Geological Survey and other Federal agencies comply with national standards of map accuracy. The standards for horizontal accuracy require that not more than 10 per cent of the tested points be in error by more than a specified distance on the map. This distance is 0.85 mm for maps published at scales larger than 1:20,000 and 0.50 mm for maps published at 1:20,000 or smaller. These limits apply to positions of such well defined points as roads, monuments, large structures, and railroads that are readily visible and can be plotted on the map within 0.25 mm of their true positions. Standards for vertical accuracy require that not more than 10 per cent of the tested elevations be in error by more than one-half of the contour interval.

Because of the prescribed standards of accuracy, topographical maps published by different agencies differ little. Some variation may be noted in format, scales, sheet boundaries, and classification and selection of planimetric details : variations primarily due to the need to meet specific requirements.

Standard topographic maps are published in quadrangles bounded by lines of latitude and longitude. Generally, topographic quadrangles cover 30 minutes, 15 minutes, 7 1/2 minutes of latitude and longitude. Scale varies with topography and contour intervals. The most common publication scales are 1 : 24,000 (the largest generally available), 1:25,000, 1:31,680, 1:48,000, 1:62,500 and 1:63,360. Coverage at 1:250,000 complied from larger scale maps is distributed by the Geological Survey for the entire country, and a new series of maps at scales of 1:50,000 and 1:100,000 is available for certain areas. The smaller scale maps are useful as the bases for general soil maps, or reference, and for schematic soil maps. Topographic maps can be used as the base for detailed mapping if recent large scale maps are available for the whole survey area.

The accuracy of standard topographic maps given them definite advantages in measuring distances and directions. The topographic pattern is helpful in understanding soil and studying drainage, irrigation and hydrology. The details on the maps relieves soil scientists of part of the task of recording the location of ground features while mapping.

As a base for soil mapping, topographic quadrangles lack the details field boundaries, isolated trees and bushes, fences and similar features that are shown on photographs. The small scale of many topographic maps is a disadvantage. The topographic maps of recent years made from aerial photographs by photogrammetric methods are much more accurate than old topographic maps which may not be accurate and may need too many revisions to be useful.

In the United States, most standard topographic maps are published by the U.S. Geological Survey. The cartographic staffs of the Soil Conservation Service receive new lists and new quadrangles as they are published and can supply information about work in progress, expected dates of completion, and the topographic mapping programme. Topographic maps needed for a soil survey can be ordered directly from the Geological Survey. Preliminary proofs or copies of manuscript material frequently can be obtained in advance of publication if the need is urgent.

Topographic maps of standard accuracy are expensive to construct and publish, but the published maps can be purchased for a small price per sheet. Besides serving as the mapping base in some areas, they are useful references.

Maps and data base requirements : The demand for natural resource data in SCS and the Federal sector has increased. In the past, these data were displayed on various base maps that generally did not meet national map accuracy standards. Soil Conservation Service (SCS) could not feasibly digitize the resource data because the use by SCS and other agencies is limited by the inaccurate bases being used. Using accurate uniform scale orthophotographs and planimetric base maps, resource data will be digitized and available for automated mapping procedures and repeated manipulation in providing various inventories and interpretative maps at great cost reduction. Repeatability of use of digitized data by SCS and other agencies, including exchanging of digitized resources data by agencies, precludes a duplication of effort in the Federal sector and results in savings in Federal mapping programmes.

Selecting Mapping Scale

The best map scale for a survey is determined by many factors. The purposes of the map are the main consideration. Soil maps in area of intensive land uses are designed for predictions about soil use, management and behaviour in relatively small areas. The scale must be large enough to permit delineation of most areas significant for such predictions. The scale does not have to be large enough to include all property lines, cultural features, works and structures for detailed plans to be plotted directly. A large scale increase the number of map sheets, the amount of joining of sheets and the cost of compilation, reproduction, publication and storage.

Most soil surveys are made at a scale of 1:24,000 or 1:12,000. A scale of 1:24,000 commonly is used for surveys in areas of less intensive land use. Scales of 1:12,000 are needed for highly detailed surveys.

Generally, the scale of mapping depends on the intensity of the soil pattern in relation to the expected intensity of soil use. The patterns of soils are very complex in many areas where potentials do not justify a mapping scale large enough to show the patterns in detail. Where the purposes of the survey area may have high value of intensive land use that justifies a scale larger than that of the rest of the area. Two publication scales can be used in such an area if the needs justify the extra costs.

Legibility of the maps is very important. Many potential users will not use maps that they cannot read easily. Figures 2 & 10 illustrates differing legibility of the same map at different scales. Map C is clearly illegible. Map B can be read with difficulty. Map A is reasonably legible. If the map is to be published at scale B, that is legible only at scale A should not be delineated.

A general idea of the smallest areas that can be shown legibly at different scales. These sizes are for isolated areas within much larger delineations. If numerous intermingled areas of the smallest size are delineated, the map will be difficult to use.

If the field sheets are made at the planned publication scale, the amount of details that should be drawn in the field is limited to that judged adequate for the purposes of the published survey. Using the publication scale also eliminates the necessity of transferring the field mapping to a different scale. If mapping scale is larger than publication scale, the surveyor should try to visualize what the map will look like at publication scale. A reducing lens can be used.

Reference Maps

Many types of maps are published by public and private agencies. They range from small scale road maps prepared by oil companies and county highway maps prepared by State highway departments to the large scale detailed maps used in city planning.

Most reference maps are designed, constructed and reproduced to meet a special purpose. Necessary details are emphasized and others are subordinated. On small scale road maps for example, highways, highway number, towns and cities, points of interest and mileages are prominently shown, drainage, railroads, pipelines, power lines and public land lines are omitted or subordinated.

Aeronautical charts: These are designed and constructed specifically for air navigation. The scale is small so that large areas can be shown on a single sheet. Ground features that are prominent from the air are emphasized in bold and simple symbols. Other features of equal importance on the ground but less noticeable from the air are subdued or omitted entirely. Elevation is shown by gradient tints. Navigational data are shown by bright overprinting.

Plats: These are prepared from public land surveys and are designed to present land survey data. They usually cover a survey unit, such as a township. The scale is large. Courses and distances, subdivisions of sections, acreage figures and other data from the survey are shown. Cultural and drainage features are reduced to a minimum and are accurate only on the survey lines.

Special purpose maps have little value as bases for detailed soil surveys. Such maps are very useful for references, however and they may be the best base maps available for surveys of remote areas. Aeronautical charts, for example, are useful for rapid small scale surveys of large areas.

Many other kinds of special maps are available for some areas. These include maps of published soil surveys, maps of geology, maps of forest or other vegetative cover, coast and harbor charts, census maps, U.S. Postal Service maps, and highway maps. County highway planning maps are available for many areas and the good references. Some State highway or transportation departments make good small to medium scale highway planning maps for internal use that can be reproduced with special permission. Maps protected by copyright cannot be reproduced without permission.

Index Maps for Field Sheets

An index map is prepared to show approximately the location of each field sheet. A useful scale is about 1:125,000. Many States have county highway maps at about this scale and many of these are good bases for preparing the index.

The mapping limits of each field sheet are drawn on index map and the field sheet number is written in each area. The preferred position for the label "Index to Field Sheets" is at the top center. Indexing by column and row makes the sheets easy to use. For example, if the northwest sheet is 1-1 (column 1, row 2) and the sheet east is 2-1 (column 2, row 1). The index accompanies the completed survey when it is submitted for map assembly.

Field and Office Activities

Preliminary Research: The soil survey party leader should arrive in the area before soil mapping begins and generally before the other party members do. This allows the party leader time to become familiar with the area, review preliminary data, investigate the major soils and their pattern of occurrence, review for the purposes of the survey, check the adequacy of the base map material and prepare a preliminary mapping legend. During the general premapping appraisal of the survey area, the party leader also assembles the information needed to schedule survey operations.

A well established principle of research is to assemble the existing information about a subject first. Time and effort are saved and costly errors are avoided if what is already known is used. The time required to findout and appraise existing information is usually small relative to the time required to compensate for failure to use the information. Even for areas about which little is thought to be known, a delligent search usually uncovers useful information. In addition, information about adjacent areas can often be applied to the survey area.

If an older soil survey has been made, it is generally the most important reference available. Soil surveys made in the United States before 1920 emphasized the character of the parent material. The maps commonly provide some of the best information available for dividing the survey area into sections within which parent material is reasonably uniform. Many soil surveys made between 1920 and 1930 provide most of the information needed to broadly characterize the area and its soils. Those made between 1930 and 1940 provide a very important part of the information needed for identifying map units. The earlier surveys are also useful for identifying map units, but they must be used in conjunction with a systematic preliminary study. It is helpful to examine mapping and example of established soil series in nearby areas that have been acqurately surveyed.

Unpublished soil surveys of scattered forms are another source of information about the area. The value of this information depends on the quality of the legend and consistency of mapping over long periods. Regardless of the quality of the legend, the scattered form mapping should not be made a part of the modern soil survey without careful field checking.

A soil survey is a study of the Geography of soil. Maps have detailed geographic information, aerial photographs, topographic maps and other maps are useful references whether or not they are used as the mapping base. Each kind of map shows features that the others do not shows any information.

Topographic maps are the best references for appraising relief for most areas.

EARTH AT A GLANCE

Composition of the Earth

Aluminium	:	0.4 %
Sulphur	:	2.7 %
Silicon	:	13.0 %
Oxygen	:	28.0 %
Calcium	:	1.2 %
Nickel	:	2.7 %
Magnesium	:	17.0 %
Iron	:	35.0 %
Total	:	100.00

Data of the Earth

Superficial Area	:	510 100 500 sq km
Land Surface	:	148 950 800 sq km
Water Surface	:	361 149 700 sq km

Great Deserts

	Area in Sq. km
Sahara, N. Africa	8,400,000
Australian, Australia	1,550,000
Arabian, Arabia	1,300,000
Gobi, Mangolia, China	1,040,000
Thar, N.W. India	260,000

CONVERSION FACTORS

A. Temperature

(*a*) Celcius to Fahrenheit 1.8, then add 32

(*b*) Fahreinheit to Celcius substract 32, then 0.55556

B. Capacity

(*a*) Pound ounces to millilitiers 28.5 militating to fluid ounces 0.0351

(*b*) Pints to liters 0.568 Litres to Pints 1.760 (2.114 for US Pints)

(*c*) Gallons to liters 4.55 Litres to Gallons 0.2193 (0.2643)

C. Weight

(*a*)	Ounces to Grams	28.3495	Gram to Ounces	0.0352
(*b*)	Pounds to Grams	455.592	Gram to Pound	0.0022
(*c*)	Pounds to Kilogram	0.4536	Kilogram to Pound	2.2046
(*d*)	Tons to Ton	1.0161	Ton to tons	0.9842

D. Length

(*a*)	Inches to cms	2.54	Cms to Inches	0.3937
(*b*)	Inches to meter	0.245	Meters to Inches	39.37
(*c*)	Feet to meter	0.3048	Meters to Feet	3.2808
(*d*)	Yards to meter	0.9144	Meters to Yard	1.0936
(*e*)	Miles to Kilometer	1.6094	Kilometer to Miles	0.6214

E. Area

(*a*)	1 sq mile to 1 sq km	2.5898	1 sq km to 1 sq mile	0.3861
(*b*)	Acre to 1 Hectare	0.4047	1 sq km to Acre	247.1
(*c*)	Acres to 1 sq km	0.004025	1 Hectare to Acre	2.471
(*d*)	Sq. feet to 1 sq meter	0.0929	1 sq meter to 1 sq hect	10.7636

F.

Name	No.	Factor	Precipe	Symbol
1 Million	1,000,000	106	Mega	M
1 Billion	1,000,000,000	109	Gaga	G

G. Basic Units

(*a*) Kelvin K Temperature
(*b*) Ampere A electric current
(*c*) Mole Md amount of substance
(*d*) Candels Cd Lumitious intensity Watt W power
(*e*) Coulomb C current
(*f*) Joule J energy
(*g*) Hertz Hz frequency
(*h*) Lux Lx illuminance (Light)
(*i*) Ohm elec. resistance
(j) Volt V elec. potential

1/3 RD LAND TO BE A FOREST ... CONCEPT 1964-65 FOREST INVENTORY SOURCE : FAO

Continental Zones	*Forested Land in Million acres*
North & Central America	2041
South America	2200
Africa	1754
Europe	356
USSR	2250
Asia & Pacific	1595
Total Global Terrain Cover :	32208 Million acres
Total Global Forest Cover :	10193 Million acres

TIPS OF SOIL CONSERVATION

Recent researches reveals that some 75 billion tons of 'Top Soil' (the fertile layer) are eroded away every year by different Geological agents and converted to wasteland. More usually, when new soil is formed from parent rocks, one cm of soil may take from 100 to 1000 years to build up. Soil is no longer a renewable resources. Only the top thin layer of the soil with humus is vital to grow food and is sustaining the entire civilization. Every year 27 mha of valuable fertile and productive land out of total 3200 mha world's potentially food producting land is being lost by desertification. Every minute a chunk of fertile agriculture land of the size of a tennis court is being converted into wasteland.

Thus, loss of soil (soil erosion) is an age old problem. To face this problem soil conservation techiniques are being applied. Soil conservation is the protection of soil against excessive loss of fertility by natural, biological, chemical or artificial means. All of us can take part in soil conservation process. There are many ways that each of us can help in solving this environmental problems at domestic level. Some of the tips of soil conservation to be followed are as follows :

- Soil should be conserved properly.
- Do not leave any patch free of plants or grasses in your premises, gardens farms.
- Do not let your garden or field soil naked and avoid the coming of it in contact with direct wind. For this, high rate of vegetation is the best solution. Polyculture should be encouraged.
- Do as much vegetation as you can in your garden farm. Because more plantation means more binding of soil. Hence soil loss will be minimized to a greater extent. Plants should be soil binding.
- Do not water in soil with a great force and in a haphazard way. It erodes the top soil layer. Use water sprinkling cans for uniform watering to plants.
- If you plan to plant on a steep slope in your premises, farm or garden, the soil can be conserved by terracing the area. It helps in slowing the rain water flowing down as water soaking will take place rather than carrying soil away. Monocotyledon plants, Casuarina plants should be planted uniformly for binding the soil.
- Avoid removing the cut grass, live it on the lawn. Cuttings serve as moisture retention mulch and a natural fertilizer. Uniform plantation of grasses should be encouraged.
- Cover the soil in your farm or garden with a layer of mulch to prevent soil erosion in the rainy season and to conserve soil moisture. Mulch can be made from grass clippings or leaf litter. Mangroves, Casuarina, Soap nut trees are best for soil aerosol.
- Compost from kitchen scarps, from poultry and cow sheds acts as organic matter, which enrich the soil naturally. Use them and introduce earthworms in the soil for best manure preparation.

- Support environmental campaigns in your state and community against cutting down of trees, like a Chipko movement. Avoid podu or shifting cultivation.
- Educate the people for aforestation programmes and Joint forest management programmes for soil conservation.

SOIL ORGANISATIONS

1. The Nature Conservancy
   1815, North Lynn,
   Arlington, Virginia, 22209, USA
2. Soil and Water Conservation Society,
   7515, Northeast, Ankeny Road,
   Ankeny, lowa, 50021, USA
3. Soil Conservation Service,
   US Department of Agriculture,
   P.O. Box : 2890
   Washington, DC, 20013, USA
4. Alliance and Environmental Education,
   51, Main Street, P.O. Box : 368
   The Plains, Virginia, 22171, USA
5. National Wildlife Federation
   1400, 16th Street, NW,
   Washington, DC, 20031, USA
6. The Wilderness Society,
   900, 17th Street, NW,
   Washington DC, 20036, USA

TEN COMMANDMENTS FOR DAILY LIFE OF LIVING

1. Care Mother Earth
2. Medicate, Exercise and Walking
3. Take Vegetarian Food
4. Eat Modrate or Loss
5. Eat 3 hours before sleep
6. Drink 3 Liters of Water per Day
7. Manage Stress & Strain
8. Check your Health regularly
9. Live Gratitude to all
10. Be smile & vibrate Love to all

LAND AREA AND POPULATION IN DIFFERENT STATES IN INDIA

States	*Capital*	*Land Area Sq. Km*	*Population*	*% to All India*
1. Andhara Pradesh	Hyderabad	275,068	66,5089,008	7.86
2. Arunachal Pradesh	Itanagar	83,743	864,559	0.10
3. Assam	Dispur	78,438	22,414,322	2.65
4. Bihar	Patna	173,877	86,374,465	10.21
5. Goa	Panaji	3,702	1,169,793	0.14
6. Gujarat	Gandhinagar	196,024	41,309,582	4.88
7. Haryana	Chandigarh	44,212,	16,463,618	1.94
8. Himachal Pradesh	Shimla	55,673	5,170,877	0.61
9. Jammu & Kashmir	Srinagar/Jammu	222,236	7,718,700	0.91
10. Karnataka	Bangalore	191,791	44,977,201	5.31
11. Kerala	Thiruvanantha-puram	38,863	29,698,518	3.44
12. Madhya Pradesh	Bhopal	443,446	66,181,170	7.82
13. Maharashtra	Mumbai	307.690	78,937,187	9.33
14. Manipur	Imphal	22,327	1,837,119	0.22
15. Meghalaya	Shillong	22,429	1,774,778	0.21
16. Mizoram	Aizawal	21,081	689,756	0.80
17. Nagaland	Kohima	16,579	1,209,546	0.14
18. Orissa	Bhubaneswar	155,707	31,659,736	3.74
19. Punjab	Chandigarh	50,362	44,005,990	5.20
21. Sikkim	Gangtok	7,096	406,457	0.05
22. Tamil Nadu	Chennai	130,058	55,858,946	6.60
23. Tripura	Agartala	10,486	2,757,205	0.33
24. Uttar Pradesh	Lucknow	294,411	139,112,287	16.44
25. West Bengal	Kolkata	88,752	68,077.965	8.04
26. India	New Delhi	3,287,263	846,302,68	-

"The Environmental Problems of Developing Countries are not the side effects of excessive Industrialisation but reflect the inadequency of Development. The rich countries may look upon Development as the cause of Environmental destruction, but to us it is one of the primary means of improving the Environment for living or providing food, water, sanitation and shelter; of making the deserts Green and the mountains Habitable. The Research and preserverance of dedicated people have given us an insight which is likely to play an important part in the shaping of our future plans. We see that however much man hankers after material goods, they can never be given him full satisfaction. Thus the higher standard of living must be achieved without alienating people from their heritage and without despoiling Nature of its Beauty, Freshness and Purity so essential to our lives."

Smt. Indira Gandhi
Prime Minister, Republic of India

UN Conference on Human Environment,
Stockholm, Sweden
June 5, 1972.

Every child deserves to grow up with water that is pure to drink, lakes that are safe for swimming, rivers that are teeming with fish. We have to act now to combat these pollution challenges with new protections to give all our children the gift of clean, safe water in the 21st century. Protect our Environment and conserve Our Natural Resources.

Baltimore, MD, USA
February 9, 1998

Mr. Bill Clinton
President
United States of America

Remarks to the Living Classroom Foundation in Baltimore, Maryland, USA announcing the Clean Water Action Plan on February 19, 1998

Acknowledgements

The author have been fortunate to have had the support and encouragement of many scientists. The author would like to extend his deepest gratitude to a number of them, without their help this book on Soil Conservation Management and Analysis would not have been possible.

Sincere thanks to Dr. S.M. Mukherjee, Former Joint Secretary, University Grants Commission, New Delhi, who helped one way or other and providing critique, moral support and technical information for success of this book. Special appreciation beyond words goes to my betraff, Mrs. Madireddi Ratnakumari, my children, Sunil, Madhavi and Siva and my grand children, Shriya, Rhea, Sai Suhaas Sai Sanjit and Sai Uanjit, Sai Darhaas, Sai Sanvita for their patience during my work on this book. My sincere thanks to the Librarians in Waukesha Public Library, USA and New Berlin Public Library, USA, during my visits to USA (2006 and 2008) for collection of literature on soils. My thanks are due to Mr. M. Eswar Kumar of EnRA for computer typing and for this book in order.

The Author heartily welcomes any criticism or suggestions regarding the contents of this book that would help to revise the Monograph and to make this more applicable.

Finally my special thanks to M/s. Daya Publishing House, Daryaganj, New Delhi for their effort beyond the call of duty in designing and illustrating the cover page and bringing the printed book in time.

Visakhapatnam, A.P., **Prof. Madireddi V. Subba Rao, Ph.D.,**

June 5th 2012 **email-D:mvsubbarao_enra@yahoo.co.in**

BIBLIOGRAPHY

1. Akroyd, TNW, 1957, Laboratory Testing in Soil Engineering. John Wiley & Sons, New York, USA, 244 pp.
2. Brady, Nyle C. 1984. The native and properties of soils (9th edition). Mac Milan Publishing Co., New York, USA, 750 pp.
3. Cook, R.L & Boyd G. Ellis (1987). Soil management. John Wiley & Sons, New York, USA, 414 pp.
4. Davidson, Donald A. (1980). Soils and land use planning. Longman, New York, USA, 364 pp.
5. Donahue Roy L., R.H. Follett & Rodney W. Tulloch (1990). Ours soils and their management (6th edition). Interstate publishers Inc., Danville, Illinois, USA, 594 pp.
6. Donahue Roy L., Miller R.W. & Shikluna C. (1983). Soils-An introduction to soils and plant growth (5th edition). Prentice Hall Inc. New Jersey, USA, 668 pp.
7. Hausenbuiller, R.L. (1985). Soil Science, Principles and Practises, (3rd edition) W.C. Brown, lowa, USA, 610 pp.
8. Knuti Leo L., David L. Williams & J.C. Hide (1984). Profitable Soil management (4th edition). Prentice Hall Inc., New Jersey, USA, 249-267.
9. Krug E.C. & Fink C.R. (1983). Acid rain on acid soils. A New Perspective Science, 221 : 520 - 525.
10. Lamba, T.W. (1951). Soil Testing for Engineering. John Wiley & Sons, New York, USA.
11. Subba Rao M.V. (1996) Soil Analysis. EnRA Publication, Visakhapatnam, India.
12. Subba Rao M.V. (1998) A Manual of Practical methods in Environmental Science. EnRA Publication, Visakhapatnam, India.
13. Tisdale Samuel L, Werner L. Nelson & James D. Beaton (1985). Soil fertility and Fertilizers (4th edition). Mac Milan Publishing Co., New York, USA, 754 pp.

Glossary

- **Accumulation :** The buildup of a persistent pesticide within the environment.
- **Aggregate :** A mass or cluster of soil particles (granules).
- **Agrochemical :** Any synthetic chemical used in modern, intensive agricultural systems, include Nitrate and Phosphate fertilizers and pesticides.
- **Agroforestry :** A type of farming in which arable crops are grown on the same land as trees, which are eventually harvested for timber.
- **Albedo :** That fraction of incoming light reflected by an object or material. Its value is always less than or equal to one.
- **Algal bloom :** A sudden increase in the amount of Algae in a river or lake, stimulated by the inputs of nutrients such as Phosphates and Nitrates.
- **Alpine :** Describing the climate, a typical of those parts of a mountain or mountain range that lie above the time but below the permanent snow line. The chief vegetation is grass and other low growing plants.
- **Arid :** Very dry region and has little vegetation (N. Africa, Pakistan, Austria, USA).
- **Avalanche :** Large mass of snow and ice, earth or rock, sliding down a mountain side.
- **Biodegradable :** Substance that can be broken down by living organisms. (Bacteria & Fungi) into its simple substance (Vs nonbiodegradable substances are glass, heavy metal & plastics).
- **Biodiversity :** An overall measure of the variety of the Earth's animal, plant and microbial species, of genetic differences within spores, and of the ecosystems that support those

species. The maintenance of biodiversity is important for ecological stability and as a resource for research (For e.g., New drugs & crops)

- **Biological Control :** The control of pests by living organisms, such as insects and fungi rather than by the use of chemicals.
- **BOD :** The amount of dissolved O_2 takenup by micro-organisms in a sample of water in order to oxidize fully any organic material in the sample.
- **Biomass :** The total mass of living organism in a given preparation or tropic level.
- **Biome :** A broad ecological community of plants and animals shaped by common patterns of vegetation and climate.
- **Biosphere :** The part of the Earth that supports life.
- **Biotechnology :** The use of living organisms to manufacture food, drug products in an industrial scale.
- **Calcicole :** Any plant that grows best on alkaline soils (Rich in Calcium in the form of limestone) Vs calcifuge.
- **Calcifuge :** Any plant that grows best on acid soils. Also known as Acidophils.
- **Canopy :** The upper level of native woodland formed by the highest branches and foliage of trees.
- **Catchment area :** A usually hilly or mountainous area which collects rainfall and channels it into a specific system of streams and rivers.
- **Cloud forest :** A type of forest that grows at high altitudes on tropical mountains, which receives constant moisture from water condensing out of the air as clouds.
- **Competitive Exclusive Principle (Gause's Principle) :** Two species having identical ecological requirements cannot coexist in the same habitat unless there is a sufficient food in that particular habitat.
- **Condensation :** The change of phase of vapour into a liquid.
- **Conservation :** The planning and management actions taken to protect and preserve the natural resources usually from pollution, over exploitation and other harmful features of human activity.
- **Coriolis effect :** The effect of the Earth's rotation on the atmosphere and on all objects on the earth's surface.

- **Deciduous :** Describes trees and shrubs that shed their leaves at the end of the growing season or during dry season to reduce the loss of water from the leaves by transportation (Evaporation). *e.g.* : Angiosperm plants - seeds are endorsed with in an avony - Now broad leaved plants-Evergreen in tropics.
- **Defoliation :** The removal of leaves, particularly from trees, through the action of drought or pests. The normal seasonal loss of leaves of deciduous plants is also known as defoliation.
- **Derelict :** Land that has been damaged by industrial processes or neglect and is unfit for any use without rehabilitating treatment.
- **Desertification :** The creation of deserts caused by climatic changes or by artificial processes.
- **Desiccation :** Drying out through the loss or unavailability of water, as in a prolonged drought. The deliberate drying out of newly filled timber seasoning is also known as desiccation.
- **Doldrums :** A calm, windless region occurring in Oceans that lie astride the equator.
- **Effluent :** Liquid sewage or industrial waste, partially or completely treated, that flows out of a treatment plant.
- **Imino :** Originally a warm current that periodically (about once every ten years) flows south along the northwestern coast of South America where the current is normally cold.
- **Endangered species :** Any plant or animal whose numbers are so few that it is at risk of becoming extinct. Officially designated endangered species are listed by the IUCN in a series of "Red Data Books".
- **Energy efficiency :** Preserving energy resources by reducing the use of energy. Profligate and inefficient energy use by industrialized countries contributes greatly to atmospheric pollution and the Greenhouse effect, when it draws on non renewable energy sources.
- **Environmental Resistance :** The restriction of population growth due to environmental factors. For e.g., Increased numbers of grazing animals leads to increased competition for between them, reducing the food supply, which prevents population growth.

- **Exotic species :** Any non native species of animal or plant that requires human intervention to survive.
- **Fallow :** Land left for season after being ploughed. Fallow land is used in some modern corporation (For eg., Podu or Shifting cultivation).
- **Fungicide :** A chemical used to control fungal diseases in plants and inorganic and organic compounds containing Sulphur which are widely used (see Pesticide).
- **Gain Hypothesis :** This Hypothesis was elaborated by the British Scientist, James Love Lock in 1970s. The idea embraced by New Age thinkers and sitters that the Earth's living and non living system form an inseparable whole, that is regulated and kept adapted for life by the activities of living organisms themselves.
- **Gashed :** A fuel for the internal combustion of engine in cars, consisting of gasoline with the addition of 10 to 20 % alcohol.
- **Genetic Resources :** An environment, such as a rainforest, that is rich in genetic diversity.
- **Global warming :** Excessive CO_2 in the atmosphere – Rises temperature.
- **Green House Effect :** The increase in the temperature of the Earth's atmosphere caused by solar radiation, absorbed by and remitted from the earth's surface, being prevented from escaping by various gases in the air. The main Green house gases are CO_2 and CFCs. (See Global warming).
- **Hard pan :** A layer of soil i.e. compacted and examined by Iron oxide, Silica, Organic material or other substances.
- **Horizon :** If a hole is dug in any well drained upland soil, there can be seen on its walls, a series of horizontal layer of soil of varying thickness, colour, texture etc. These layers are Horizons.
- **Hurricane :** A revolving storm in tropical regions (also known as cyclone or typhoon).
- **Hydroponics :** A method of cultivating plants instant soil. The plants are grown in peat or gravel, through which a solution of artificial nutrients is circulated.
- **Iceberg :** A mass of the floating on the sea, about 80% of which is submerged. Ice bergs are formed from glaciers or Ice barriers.

- **Indicator Species :** Plants or animals whose presence or absence in an area indicates certain environmental conditions, such as soil type or low levels of dissolved oxygen in rivers. Many plants prefer either alkaline or acid soil. Certain trees are found only in soils with high concentration of Aluminium. Some lichens are sensitive to Sulphur dioxide in the air.
- **Integrated control :** A system of pest management that uses the principles of biological control together with general pesticides and other practices such as crop rotation (See Organic farming).
- **Intensive farming :** Maximizing crop yields and livestock produce, using efficient machinery.
- **Irruption :** The sudden emigration of a population (Practical or total) usually in response to over crowding or food shortage.
- **Leaching :** Removal of dissolved materials by water.
- **Marl :** Soft, earthy material consisting of a large amount of $CaCo_3$ with various impurities, alkaline in reaction.
- **Marginal land :** Poor quality of agricultural land that is likely to yield a low return.
- **Microbeam :** An ecological community of organisms that occupies a relatively small area, such as a woodland, lake or hedger.
- **Microclimate :** The climate of a very local area. Microelements play a significant role in Agriculture and Horticulture because different crops require different growing conditions.
- **Mudflow :** A tract of low muddy land composed of five sites found near estuarine regions that are covered at high tide and exposed at low tide.
- **Nomadism :** A system of livestock farming in which animals are taken to fresh pastures in different locations. Large animals held may lead to overgrazing and desertification. Nomadism is under threat due to increasing enclosure of land.
- **Opencast Mining :** The extraction of minerals, particularly metal ores and coal by excavating.
- **Pedology :** The study of distribution and morphology of soil. Pedology is an important aspect of agricultural science.
- **Pedologist :** Soil Scientist.
- **Permafrost :** Ground that is permanently frozen. It covers approximately 26% of the world's land surface in alpine areas.

- **Persistant :** Describing pesticides that are resistance to decay in the environment. They may accumulate in the environment. Persistent pesticides may also give rise to resistance strains of pests. (See Resistance).
- **Pioneer species :** Species that are the first to colorist and thrive in new areas. For eg : Coal tips areas.
- **Plagioclimax :** An artificial climax community that forms following the clearance of existing vegetation. Exs : Health land and Down land.
- **Podozol :** A type of soil associated with plants that favour acid conditions.
- **Pollution :** The hazardous effects on the environment caused by products of human activity. Pollutants may enter the food chain and be passed on from one organism to another causes additional damaging developments, such as acid rain.
- **Profile :** The succession of Horizons from the surface down to and including the underlying rock or other parent material.
- **Puddle :** To destroy the granular structure of a soil.
- **Quartz :** Hard, tough mineral made up of the elements of Sillicon and Oxygen.
- **Rainforest :** Dense forest found in equational regions where the climate is hot and wet.
- **Reclamation :** The conversion of swarms, marsh, desert or other wastelands into land suitable for cultivation or settlement.
- **Ruderal :** Describing plants that inhabit rubbish heaps, wasteland, old fields or waysides.
- **Seed bank :** A store of seeds in the soil that may lie dormant for years. Often the seeds germinate only when the soil has been disturbed, bringing them into contact with an environmental trigger, such as change in troupe, Co_2, height of water. The term is also used for a store of seeds, that is kept in the laboratory for use to maintain genetic resources and diversity.
- **Soil Capability Map :** Map of the different soils of an area, used to decide what crops can be grown in each kind of soil.
- **Soil Conservation Service (USA) :** Administers programmes to develop and conserve soil and water resources, offers technical association on Agricultural pollution control and Environmental improvement projects (Agricultural Dept.)

- **Sub Soil :** Soil below A-Horizon or Top Soil. The part of the soil below top soil and above the under lying rock or other parent material.
- **Slash-and-burn :** A simple agriculture practice in which natural vegetation is cut and burned, and clearing, then formed for a few years until the soil loses its fertility, where upon farmers move on and leave the area to another region.
- **Soil :** A loss covering of broken rocky material and decaying organic matter overlying the earth's surface.
- **Tallith :** A general term indicating the physical condition and workability of the soil.
- **Top Soil :** The surface layer of soil i.e., the living soil containing organic matter and turning with microorganisms.
- **Tornado :** An extremely violent revolving storm with swilling, funnel shaped curves, caused by a rising column of warm our propelled by strong wind. Tornados move with wind speeds of 160-480 km/h, destroying everything in their path.
- **Tundra :** Treeless arctic and alpine regions of high latitude only. Grosser, sedges, mossier and lechers are present in Tundra.
- **Typhoon :** A violently revolving strong (see Hurricane) UV Radiation, Electronic Magnetic radiation from the Sun, in the wavelength range from 400 to 10 nanometers. Harmful to living organisms.
- **Weathering :** Process by which rocks, which are broken into small particles by the action of weather.
- **Watershed :** A divide separating two drainage basins (catchment area) usually a sides of high ground.
- **Zonation :** The occurrence of all area of distant bounds of vegetation, each in its own characteristic dominant species *e.g.* : Seashore & Mountain sides.

> *"Protect our Environment and Conserve Our Natural Resources".*
>
> —*EnRA.*

❑❑❑

www.ingramcontent.com/pod-product-compliance
Ingram Content Group UK Ltd.
Pitfield, Milton Keynes, MK11 3LW, UK
UKHW020207290926
15470UKWH00001B/38

9 789351 301233